Brain, Mind, and the Narrative Imagination

*Christopher Comer and
Ashley Taggart*

BLOOMSBURY ACADEMIC
LONDON • NEW YORK • OXFORD • NEW DELHI • SYDNEY

BLOOMSBURY ACADEMIC
Bloomsbury Publishing Plc
50 Bedford Square, London, WC1B 3DP, UK
1385 Broadway, New York, NY 10018, USA
29 Earlsfort Terrace, Dublin 2, Ireland

BLOOMSBURY, BLOOMSBURY ACADEMIC and the Diana logo
are trademarks of Bloomsbury Publishing Plc

First published in Great Britain 2021

Cover design: BuroBlikgoed.nl
Illustration: © Shutterstock.com

A catalogue record for this book is available from the British Library.

Library of Congress Cataloging-in-Publication Data
Names: Comer, Christopher Mark, 1952- author. | Taggart, Ashley, author.
Title: Brain, mind and the narrative imagination / Christopher Comer and Ashley Taggart.
Description: London ; New York : Bloomsbury Academic, 2021. |
Includes bibliographical references and index.
Identifiers: LCCN 2020038550 (print) | LCCN 2020038551 (ebook) | ISBN
9781350127807 (hardback) | ISBN 9781350127791 (paperback) | ISBN
9781350127814 (epub) | ISBN 9781350127821 (ebook)
Subjects: LCSH: Narration (Rhetoric) | Neurosciences and the humanities. | Imagination.
Classification: LCC PN212 .C59 2021 (print) | LCC PN212 (ebook) |
DDC 808/.036–dc23
LC record available at https://lccn.loc.gov/2020038550
LC ebook record available at https://lccn.loc.gov/2020038551

ISBN: HB: 978-1-3501-2780-7
          PB: 978-1-3501-2779-1
          ePDF: 978-1-3501-2782-1
          eBook: 978-1-3501-2781-4

Typeset by Integra Software Services Pvt. Ltd.
Printed and bound in Great Britain

To find out more about our authors and books visit www.bloomsbury.com
and sign up for our newsletters.

CONTENTS

FIGURES

PREFACE

Our interest in science and the narrative imagination began with two deceptively simple questions: how do stories, as told aloud or written, have an impact on us, and what can science, in particular the burgeoning field of neuroscience, bring to our understanding of this process?

For a very long time, as we will show, the arts and sciences were viewed and pursued in isolation, as "two cultures" with little common ground between them, and radically incompatible aims, methods, and criteria of excellence. But this "cold war" is breaking down. A dialogue, if not quite *détente*, has broken out. For anyone interested in narrative, we live in promising times. Our increasing grasp of what goes on in the brain not only turns out to be relevant to the experience of narrative, but is now taken seriously in fiction, essays, and commentaries by those who produce it, storytellers in every genre—a sure sign that at the very least "something is afoot." Even though the two cultures may preserve different "languages," there is now considerable two-way traffic between them. Empirically derived ideas about structure, form, and even the roots of narrative thought and imagination have at least started to provide common ground which is of equal interest to the novelist, narratologist, and neuroscientist.

In order to show how the iron curtain between the narrative arts and science is crumbling, inclusivity has to be a key principle. Hence, the term *narrative imagination* is not restricted here to the "imaginative act" of conceiving narratives or, in fiction, to characters, scenes, and plots for oral, written, or enacted delivery. For a start, studies of the brain offer perspectives that extend well beyond the moment of creativity, fascinating though that is. They have now made great strides on the question of what happens to a solitary reader, or an audience, when receiving and internalizing narratives of all sorts. Then there is the "return traffic"—less congested perhaps—of those who want to come over and survey the concepts of science, and the scientific process with a narrative eye.

Our exploratory journey into narrative imagination was initially prompted by the development of a university course designed to encourage students from both fields to cross the apparently vast divide between them. In effect, we wanted to help them take a walk across campus, between faculties—itself an interesting word in context: a short distance on the map, perhaps, but dauntingly cosmic in terms of understanding, and even respect. It is all too often an untrodden path, beset by the demons of aversion, incomprehension, and misunderstanding. If those in the sciences cannot see the point of sitting in seminar rooms endlessly discussing often idiosyncratic reactions to stories, plays, and films with theoretical jargon, those in the arts are equally repelled by the hegemonic claims on objectivity, verifiability, and progress made by science; its hunger for research funds; and the daunting impedimenta of labs, scans, and technical jargon. It is not just a question of *what* is being pursued on either side, but *why*. For those in the arts, the most comprehensive grasp of neuromodulators and synapses will never bring us one jot closer to the sense of joyous transcendence produced by a poem like Philip Larkin's *High Windows*, or the sublime interconnectedness of *Middlemarch*.

On the other side of campus, the achievements of brain science barely need enumeration. Every month new journal articles and books appear, which seem to fill in another blank in our general understanding of the brain, but—and this is the interesting thing—often by focusing on issues that are the lifeblood of the arts. The physical processes behind memory, imagination, feeling, and even creativity are relentlessly pursued by scans and research studies. Yet, for many in the humanities, this effort all too often appears worthy but wrong-headed, like watching a game with a sports nerd who is so obsessed with batting averages they have forgotten to enjoy the spectacle. What brings us to the stadium in the first place is somehow misplaced: that vicarious ebb and flow of anticipation and anxiety familiar to every sports fan, buried under analysis and statistical probability. The magic is lost.

But there is more than this going on. After all, there is a whole production line of critical analysis in the narrative arts too, wielding the tools of structural breakdown, close-reading, exegesis, narratology, and stylistics, to get closer to the storyteller's cabinet of tricks. It appears, then, that in this camp too, concerns about turning creative diamonds into critical ash can be set aside when it suits.

When confronting something as fundamental and all-pervasive, something so distinctively human, as narrative, it is easy to become caught up in skirmishes, internal rivalries, and power struggles. We can become overwhelmed by the details, small advances, or fixated on the barriers between the "two cultures," fortified over many years. To be sure, there remain many differences, not just of methodology, but of outlook and intent. But to get a clearer view it is useful to take a step back—*reculer pour mieux sauter*—and forego complexity for simplicity, asking "what is the state of play right now, and what are the motivations behind it?" When we do this, it's apparent that, lurking behind all the residual distrust and mutual incomprehension lies anxiety and enigma.

The anxiety, from the humanist side, is the fear of annexation, of displacement, and worse, irrelevance. We can hear strains of this anxiety throughout the arts—from critics and practitioners alike. What will be left for the novelist when the neuroscientists are done? For the critic? In the era of ever-more sensitive brain scans, talk of "conscious" machines, and story algorithms, speculation on the death of narrative comes to seem less and less like some fanciful conceit. Reading the recent rash of so-called neuronovels, you sometimes get the sense of the writer incorporating their own death notice into the story, in the hope that this will avert the actual event: narrative becomes its own elegy. Certainly, we can hear this doleful refrain, intrusive and discordant, in the work of Ian McEwan and E. L. Doctorow, two writers we will look at later.

The enigma is of course the mind-body problem, which casts an Easter Island-like shadow over the endeavors of scientist and storyteller alike. To recast it slightly, the big question here, whether acknowledged or not by either party (and it is often simply disregarded), is whether the mind is the brain. Or crucially, whether mind is the processes of the brain—its scintillant myriad interconnections—or whether it is something more. To return to our earlier analogy, the philosopher Gilbert Ryle compared "dualists" searching for an essential "self" or inner mind distinct from the body, as being like someone given a tour of campus, shown the library, labs, seminar rooms, sports fields, and lecture theaters, who then asks at the end of it all: "But where is the University?" His point being that all of these comprise the university, in much the same way as our behavior, predispositions, and physical attributes comprise the self. Or put another way, mind is what the brain does.

Despite recent efforts to introduce ideas of, for example, embodied linguistics, brain lateralization, and affective cognition into the humanistic study of narrative, and despite the best efforts of storytellers to assimilate what is emerging from new brain studies, it is fair to say that the mind-body problem remains a major contributing factor to mistrust and misunderstanding in this field. If, as Ryle famously said, mind is the ghost in the machine, then it remains a stubborn truth that for many scientists, the literary study of narrative is little more than chasing such ghosts. Meanwhile, for many in the arts, scientists display a blinkered obsession with mechanics, the minutiae of the brain's "wet stuff," to the exclusion of all else.

The exploration we undertake here is of course dictated by our own prejudices, preferences, and competencies, so is by its very nature partial. The terrain we seek to survey does not stand still. Luckily, we are not traveling blind, but have recourse to the charts of those who have preceded us. After all, although every map is totalizing, and to that extent idealized and aspirational, most maps tell us something. And, if we are lacking a "grand narrative," which will definitively bring together the "ghost" and the "machine," or indeed unite the humanistic and scientific perspectives, it is perhaps better simply to admit this, acknowledge the constraint, show that progress has been made, and get on with the task in hand. Yet, ignoring the elephant in the room won't make it disappear. Some stance must be taken. Our own attitude to the mind-body problem, the stance we have adopted, is a version of what is known as a "Type" theory: the assumption, based on economy of entities, as well as practicability, that mental states are, as a matter of contingent fact, identical with brain states. If this amounts to a working hypothesis, rather than a solution, it is enough to guide our course.

It is an appropriate time for a multidisciplinary look at narratives and imagination. Ingenious experiments are being devised and innovative technologies casting new light on some venerable debates about what it means, for instance, to "hear" a character speak, "feel" their emotion, visualize a fictional scene, or experience a particularized inner "storyworld." Literary critics and writers of fiction are turning to face the oncoming tide of scientific research encroaching on topics they once thought of as their preserve; whereas scientists are increasingly facing up to the role of story, metaphor, visual projection, and imagination in their own hypothesis generation, theory construction, and promulgation.

Behind what can seem like arcane academic turf wars about methodology, verifiability, and applicability lie psychological phenomena so banalized, to which we have become so deeply habituated, that they seem barely worthy of comment. Take our insatiable hunger for narrative. It is so much part of the habitual background of our ordinary life that it takes the extraordinary to cast it into relief. At the time of writing, during the COVID-19 lockdown, Netflix has seen a sudden spike in subscriptions adding nearly 16 million new viewers in the first quarter of 2020; news sources are accessed obsessively, and report, amongst the dire headlines, the fears of those who may not now get their regular "fix" of soap-operas; virtual book clubs are thriving, and in local neighborhoods people are leaving out tubs of hard copy books on a take-and-replace basis.

There are practical reasons for some of this: more time to fill, the need to find out what's happening, social isolation, and so on. But behind that lie two opposed drives. The urge for immersion in a fictional world, inner voyaging, even escapism, and the urge to confront the world at large—engage with the source of threat. These drives are manifested in most of us to a greater or lesser degree, often alternating within one person, from one moment to the next. We want to be taken out of ourselves even as we want to know where we stand.

Narrative answers both these fundamental drives, and it is perhaps only in "strange times" that we might start to seriously query the source of this daily preoccupation with creating and consuming narratives which dominate much of our waking, and some of our sleeping hours. From the briefings of governments to "human interest" stories about the socially distanced tittle-tattle of dog-walkers: narrative in all its shapes is the answer we resort to and are primed for, promising in various measure, consolation, and comprehension.

As with any other mental feature, one way to analyze narrative processing is through the creation of increasingly accurate maps of the brain. Neuroscience has now accumulated insights about mental life that we did not possess a decade ago. The very description of brain structure—its "atlas"—is undergoing radical revision, moving from a bewildering array of Latinate locations, to an interacting set of "large-scale intrinsic networks." We are finally at a point where the observation of our brain in action can be made under more natural or "ecological" conditions, allowing complex behaviors such as the comprehension of speech or writing to be studied. This information is emerging not only from brain scan techniques—on which popular reports tend to focus—but from an array of other methods that can even tell us about what single neurons are doing, or what happens when they are exposed to narrative stimulation. Topics at the very heart of narrative, such as point-of-view, focalization, empathy, and "transport," are now appreciated in greater detail than ever before. Processes that would have been thought unworthy or untouchable by science (and—imagine it—imagination was one) are now being pursued with a sense of excitement and real possibility.

Of course, the very suggestion of an empirical approach to the humanities may sound misplaced to some. With regard to the interpretative reception of literature this may be so—there is an irreducibly subjective and evaluative element to it. Yet the humanities-led study of narratives has itself generated ever more focused theories about how discourse "works," spawning ideas that feed back into cognitive science and theoretical and applied psychology. Neuroscience has added to this development with ideas on the relationship between memory and imagination, embodied cognition, and models of how culture influences mind and brain.

At one level, this recent convergence should not be surprising. As neurobiologist Paul Grobstein (2005, 2007) has pointed out, science itself is necessarily engaged in a narrative enterprise, performed in a spirit of open-ended exploration and skepticism. This is yet another indication of our natural propensity to tell stories: it is "the brain's way." Yet often, it is necessary to turn to new maps to make sense of what we find. New ways of framing enduring questions about imagination do not come only from literary study or from neuroscience *per se*, but from other fields too, such as anthropology and philosophy. We will try to accommodate some of these views along the way in this book. A pioneer of cognitive literary thinking, Wolfgang Iser, once described imagination as the essential pathway between sensation and action. This is as true of science as the arts.

The Organization of This Text

The defining principle behind this book is multidisciplinarity. We turn to many disparate fields and disciplines in the course of our journey through what might sometimes appear an impenetrable landscape. Yet we are aware, too, that sometimes these shifts in focus risk disorienting the reader. At times the resolution will be fine, prioritizing brain networks and neurons; at others, the narratological features of metaphor, authorial choice, and intertextuality will dominate. To obtain the best overview, we will shift from the figural to the literal, from the symbolic to the microscopic, as occasion demands. The temporal scale of our enquiry encompasses both neural activity taking place in milliseconds, and the "deep time" of evolutionary emergence. Perhaps the best way to regard this is as a switching between a series of maps. And if on our travels we were to come across an ancient oak forest, a geological, ecological, or climatic map might initially seem of little relevance either to its existence or to the sense of wonder it inspires; but a little thought might show otherwise.

Throughout, we emphasize what brain science can bring, now, to the discussion of narrative imagination. But to place recent insights in context there is also a need to historicize some key features along the way. This begins in the Introduction by pointing out that the disconnect between the cultural domains, between what might be called the topographical maps of science, seeking exhaustiveness and objectivity, and the more subjective topological maps of the critical Humanities, is often more apparent than real. Mercator, whose revolutionary atlas has graced schoolrooms for generations, relied on the anecdotes of travelers and explorers, even as he devised the mathematical projections which allowed him to complete his task.

We can become fixated on our maps, entranced by them to the point that we come to believe the one in front of us, and only this one, represents actuality. It is crucial, therefore, to remember the warning of Alfred Korzybsy that the map is not the territory. A truly exhaustive map on a 1:1 scale, although an idea toyed with by Jorge Louis Borges, in the satirical *On Exactitude in Science*, and Lewis Carroll before him, is a logical and theoretical impossibility. It would *be* the world it is attempting to depict. Each map is a unique intersection of knowledge and living, governed by intention, scale, and range. Classical painters and sculptors used to sign their works "faciebat"—"was being made, or is unfinished," in the awareness that perfection was forever elusive; in a similar way, the totally inclusive map is a chimera. Yet exploration and cartography are no less fascinating for the acceptance that they can never be complete.

Ancient maps used to fancifully include mythological creatures to indicate they had reached the limits of their competence. We too will encounter our fair share of sea-serpents, dragons and hippogryphs, dead-ends, and false trails; but far from invalidating the entire enterprise, this merely affords a glimpse of further horizons.

Part I considers why narratives matter, looks at some oral and written forms, their pervasive emergence across cultures and time. It reviews their unique intersubjective capacity, which allows us to understand and even vicariously re-enact the inner experiences of others. Across the range of genre and form, narratives are multimodal: stories evoke not only our inner voice, but also our inner eye and our bodily senses. We start with the basic question of how storytellers produce these effects. Drawing especially on the insights of Elaine Scarry, we examine how readers can be instructed, provoked, or enticed into summoning up a multisensory storyworld, how writers of fiction can orchestrate our imagination. As we will show, there is abundant convergence here with the recent neuroscience of imagery and affect, parallels with the literary use of point-of-view and focalization, and the psychological phenomenon of "transport."

Part II has us turning to a very different map, taking us into the physical cartography of the brain. It delves deep into what Ian McEwan calls "the wet stuff," and in some ways is analogous to a breakdown of our basic equipment, a stop at an "outfitter" for an inner journey that will venture deeper into the "three-pound universe" of the brain. This inventory provides a neural basis for later discussion of themes that become central: "embodiment," the emergence of cognitive functions from cell networks, and sensorimotor integration, that malleable "kit" we have all inherited for speech and vision, and which can, *mutatis mutandis*, become oriented toward reading. Although challenging for some, we believe that the science can and should be addressed. It is, after all, another form of storytelling.

Any discussion of narrative imagination has, ultimately, to return to words—voiced or written. **Part III** travels from the emergence of language in preliterate speech and gestures, through the individual's development of reading and writing skills, into the recent emphasis on how culture organizes and re-organizes our literate brains. Yet narratives are not composed only of words, but of sentences and paragraphs, sequential events located in space and time, and, in the case of fiction, "storyworlds." Recognition of this fact leads naturally into consideration of narrative-induced images, tantalizing recent research on the brain's "semantic atlas," and our use of language in meaning-making. To use Iris Murdoch's metaphor, we cast language like a net over the world, only to find ourselves ensnared by our own creation. We are all *Under the Net*. All the more reason to step back and end the part with a look at the fundamental ability to comprehend discourse at all—a process which mobilizes memory in pursuit of imagination.

In our journey into narrative creation and reception, we will often consult charts which relate to theme, character, and emotional response alongside those of brain states. Such shifts should not be seen as just an analytical contrivance. After all, in daily life we are quite comfortable "toggling" between descriptive, categorical, and perceptual frames depending on circumstance. We have no trouble with the knowledge that a bolt of lightning is an electrical discharge, even as we perceive it as a jagged, incandescent flash, and feel it as an awesome natural event. Not either, but also. The physical account is as valid as the phenomenological one—depending on what we are looking for. Every map, every description, is an example of what Peter Turchi calls a structure-defining intention.

To underscore this fact, the necessary partiality of our vision depending on the demands of the medium we deploy and the goals we seek, we conclude in **Part IV** with linked chapters on literature and neuroscience, which touch on common themes of emotion, sociality, and the role of narrative in our sense of self. Through a more detailed "field survey" of two highly topical novels, we hope to illustrate how neuroscientific research is not just infiltrating but also informing the work of professional narrators. These works by E. L. Doctorow and Ian McEwan are fictional outcrops, which allow us to take a broader view of the terrain already covered, ranging from cognition to feelings, and perhaps also to suggest some ways ahead. Inevitably, this final part also returns to the schism between mind and brain that bedevils any attempt to "know thyself."

The works of literature, critical analysis, and theory relating to our topic are vast, no less so than the primary research reports, reviews, and meta-analyses from the brain sciences. Yet the sometimes-daunting jargon and the perceived barrier between science and the arts mean that the number of people who are aware of what's been happening in the territory where the study of narrative and brain science overlap is relatively small. A primary aim of this book is to expand that number.

There are many findings of interest in this area that cannot be given the attention they deserve: our path through the borderlands between the two domains has been just that—one path among many. As mentioned, our approach is "quasi-historical." This has certain advantages, not least that it helps to create a framework for laying out narrative theories and scientific hypotheses in terms both of lineage and of the succession of challenges that have been applied to them over the years. Where possible, we emphasize observations that have proven robust and generative.

The overall goal is to convey some of the perils, but also the promise, of venturing into the terrain between stories and the minds that produce and enjoy them. Today, we are relentlessly exposed in the press to the latest revelations provided by brain scans, but seldom hear what the underlying cellular processes are telling us. So we have made sure to include both. One trap we were determined to avoid was that of reducing the person, or even personal traits, to bits of brain, or neurons. Again, scale, projection, scope, and intent are key—and, we might add, respect. Levels of analysis and description matter a great deal, reflecting the apparently fragmented state of our understanding. Neuroscience has much to say about information processing, but is less forthcoming on questions of personal identity, subjectivity, interpretative diversity, or the act of narration itself. The chapters below which deal with brain structures, neural networks, and synaptic signaling may seem light years removed from the virtual reality experience accorded by fiction, drama, or even a compelling anecdote. Humanists might make a more skeptical response. "What has any of this to do with 'the stuff of thought'" in Stephen Pinker's phrase?

Some patience is therefore required from the reader in our exploration of common ground. We are highly conscious that point of view, and perspective, whether actual or fictional, cannot simply be reduced to coordinates, wonder to molecules, or memory to proteins. Location is not merely a matter of space, but of place, too. But finally, our response to Pinker's question "What has any of this to do with 'the stuff of thought'?" has to be: quite a lot, as it turns out.

ACKNOWLEDGMENTS

Those who helped us logistically—in Chicago, Dublin, and Missoula—especially with the university course we taught nearly every other year in Dublin since 2004 include Joan Gillespie, Mary Dwyer, Diane Pecknold, and Marja Unkuri-Chaudhry. Colleagues who generously collaborated on components of the course include Julie Anne Stephens, Tom Halpin, and Patrick Sutton. Additional technical help and assistance in finding "space" to write came from Perry Brown, Royce Engstrom, Jenny McNulty, Jared Sheffield, and Kitte Robbins. Friends, colleagues, students, and experts have discussed ideas on brain, mind, and narrative imagination with us over the years, answered questions, reviewed drafts of chapters, or made important suggestions: Jill Bergman, Charles Birtwistle, Conor Carroll, Richard Drake, Stanley Fish, Michael Fisher, Gerald Graff, Paul Grobstein, Charlie Gross, Bob Hausmann, Eamon Jordan, Ashby Kinch, Peggy Kuhr, Steven Matthews, Ellen McCulloch-Lovell, John Miller, Michael O'Dea, Madison Padilla, Marya Schechtman, Jeremy Smith, Tully Thibeau, Chris Thiem, Kelley Willett, and Jennifer Wiley. We thank them all for their contributions. Along the way, editorial advice from David Avital, Ben Doyle, and Lucy Brown at Bloomsbury was crucial, and several anonymous readers provided valuable suggestions from which the text has surely benefited.

There are, of course, some we wish to acknowledge for personal contributions.

AT: I'd like to thank my wife, Amanda, and son, Luke, for their interest, tolerance, and encouragement throughout the process. Also, my Mum, Kathleen, who encouraged and put up with me (and sometimes put me up) during the writing. This book is dedicated to them. CC: Many, many thanks to Linda Gazzola and Nico for considerable patience and emotional support. This is dedicated to them, and to Colin, Sarah, Joseph, and Patrick; and to Veronica Comer Serio who bravely rallied from the effects of a stroke—in no small way—through reading narrative literature.

Publication of this book was supported in part by a grant from the Baldridge Book Subvention fund through the Humanities Institute of the College of Humanities and Sciences at the University of Montana. We are very happy to acknowledge this help.

ABBREVIATIONS

3D	three-dimensional
ACC	anterior cingulate cortex
ANS	autonomic nervous system
ATL	anterior temporal lobes
BCE	before the common era
cc	cubic centimeters
CNS	central nervous system
CT scan	computed tomography (essentially equivalent to MRI)
DMN	default mode network
DNA	deoxyribonucleic acid
EEG	electroencephalogram
FFA	fusiform face area
fMRI	functional magnetic resonance imaging (a scan for neural activity)
IT	inferotemporal cortex
LH	left hemisphere (of the cerebral cortex)
MEG	magnetoencephalograpy
MRI	magnetic resonance imaging (a scan for brain structure)
ms	milliseconds (1/1,000 second)

MTL   medial temporal lobe

PCC   posterior cingulate cortex

PET   positron emission tomography

PFC   prefrontal cortex (preceded by d, v, or vm: it indicates the
         dorsal, ventral, or ventromedial regions thereof)

RH   right hemisphere (of the cerebral cortex)

RMET   reading the mind in the eyes test

TMS   transcranial magnetic stimulation

VWFA   visual word form area

WEIRD   White, Educated, Industrialized, Rich, Democratic

Introduction: Back to the Future

To reach the exciting new territories currently opening up between brain science and narrative, we must first look to the past and the figure of I. A. Richards. A seminal literary critic, he stimulated and in many ways anticipated the empirical study of literature in the twentieth century, advocating a psychological approach to the literary mind. As will become clear, he was thinking with and beyond the science of his era, and consideration of his ideas helps to put subsequent interactions at the border between narrative and science into perspective.

Between 1924 and 1929, Richards, then a lecturer at Cambridge University, published two books that greatly influenced our broader perceptions of literature. The first was *Principles of Literary Criticism*, followed five years later by *Practical Criticism*. These works established an approach to understanding literature and literary meaning that was new at the time, and which evolved into an important critical approach for decades.

Richards entered the scene at a time when the teaching of the subject was still linked to grammar and rhetoric. However, he was among those arguing for it to be studied in its own right, and indeed was there at the founding of the School of Literature at Cambridge. The approach he advocated was characterized more by a concern with the "close reading" of texts than with historical and biographical factors. This relatively decontextualized attitude became a hallmark of the group that came to be known as the Cambridge Critics: Richards and his students William Empson and F. R. Leavis. Over time, this morphed into the so-called New Criticism, but, crucially, science—and in particular neuroscience—also became involved at this vital juncture.

In no small part, this was due to an extraordinary proposal made by Richards in *The Principles of Literary Criticism* (1924). Here, he asserted that, to establish an effective critical stance, one had to base it upon a view of the mind, believing that a good critic must be adept at identifying and relating to the mental state created by a work of art. Psychology, he said, "even in its present conjectural state, has a direct bearing" (114). He clearly felt that, to understand our responses to literature, it was preferable to have a conjectural account of mental phenomena than none at all. What is truly striking is that his description of what was required of the critic was phrased in terms that would not sound out of place in a neuroscience discussion today, arguing, for instance, that we have need for "a schema of the mental events which make up the experience of, 'looking at' a picture or 'reading' a poem" (114).

If you retrieve the original edition, you will notice something even more surprising. Richards gives us a very highly delineated description, a compact conjectural map, of the mental steps involved in reading and interpreting a poem. His thinking was broad, encompassing the physiological processes of reading and the psychological processes of reader response. Moreover, he was clearly alive to the fact that his model had implications extending far beyond an engagement with poetry.

What he chose to deliver was not just a humanist's description of narrative comprehension, but also a detailed graphical model of the type you might expect to see in a science journal (Figure 0.1). This model was, in his words, "a diagram, or hieroglyph ... intended as a convenience, ... provided that its limitations are clearly recognized. The spatial relations of the parts of the diagram, for instance, are not intended to stand for spatial relations between parts of what is represented" (117), a reminder that it's more topological than topographic.

As you can see, the diagram suggests that images of words on a page are projected by way of the eye to a system of repeating modular (i.e., largely self-contained) units in the brain that perform sequential operations in response to them, for example, evoking images, or emotions. While Richards stresses this sketch was not meant to be "a picture of the nervous system," it was clearly inspired by information about neural architecture. The elements within his modular units are reminiscent of nerve cells. Indeed, the whole structure is extremely suggestive of what was then known about the functional organization of the outer layers of the cerebrum: the cortex (for an outline of brain anatomy, see Chapter 3).

Even more remarkable, the model shows distinct mental operations occurring at six different levels and indeed there are known to be six structural/functional layers in the cerebral cortex (even designated I–VI as in Richards' model). However, as modern neuroscience sees it, information would not simply flow through the layers as implied (nor, as you will see, is the cortex the only level of the brain involved). All the same, his vertical structural motif is highly suggestive of the basic functional unit of the cortex, the cortical column, which was not even discovered until the 1950s. Furthermore, the model shows lateral connections between the vertical modules and again, current research indicates that there are just such lateral interactions. Rather than any nod to antiquity, Richards' so-called hieroglyph turned out to be highly prescient.

Richards' model suggests that a line of text is digested for meaning one word at a time. Once more, this proves to be highly accurate: a large body of work on reading shows that a sentence is not comprehended all at once

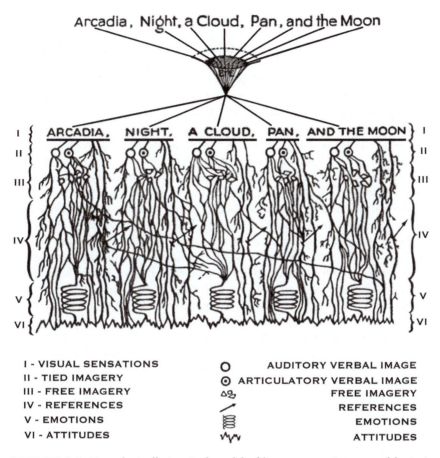

I - VISUAL SENSATIONS   O   AUDITORY VERBAL IMAGE
II - TIED IMAGERY   ⊙ ARTICULATORY VERBAL IMAGE
III - FREE IMAGERY   FREE IMAGERY
IV - REFERENCES   REFERENCES
V - EMOTIONS   EMOTIONS
VI - ATTITUDES   ATTITUDES

FIGURE 0.1 *Neurologically inspired model of literary reception created by I. A. Richards. This appears in Chapter XVI, in the first edition of Principles of Literary Criticism. For a discussion of its origin and significance see the text. From: Richards ([1924] 2001) Principles of Literary Criticism, Figure on p. 107, Routledge Press, used with permission.*

or even in a smooth stream, rather, the eyes make a series of jumps, called *saccades*, across the lines of text. As they alight briefly, they acquire the image of a group of letters roughly equal to one word and then immediately jump further along, pausing to sample another word or two. During reading, this process is repeated again and again. That such eye movements underlie comprehension has been known for more than 100 years, from the work of psychologists such as E. B. Huey (1908). Clearly, Richards was incorporating the best available experimental knowledge of the time. Even now, the complexities of eye movement continue to inform our understanding of reading dynamics, cognition, and literacy (e.g., Rayner 1998, Rayner and Reichle 2010, Walczyk et al. 2014).

Perhaps most striking here are the mental functions specified in Richards' model which distinguish a set of steps that comprise our cognitive reactions to the text. The brain first responds to the visual sensation of clusters of letters (layer I). This is followed by the elicitation of imagery in multiple modalities (II and III) and stored associations, triggered by words or phrases (IV). Layers II–IV might best be interpreted as calling up things remembered or imagined, whereas the final layers (V and VI) relate to the evocation of emotions and attitudes.

The key point is that Richards' educated hunch was right. He correctly presumed that the initial processing of text depends upon word images being recognized, followed closely by the engagement of memory and emotion. Such a model, in which the serial processing of words blends with the early involvement of memory and emotion, is consistent with current psychology and neuroscience (e.g., Marinkovic 2004, Hagoort 2008, Oatley 2011).

As mentioned, Richards' innovative mental framework also carries implications for how we should begin to evaluate the distinctive qualities of stories. Following his lead, we will focus mostly on the basic operations outlined here—the textual evocation of imagery, memory, and emotion— as landmarks for exploring the neurocognitive understanding of narrative. While some time will be given to oral narrative, which after all long predates anything written down, more of our attention will be given to textual sources as a matter of practicability, and much of that, to the specialized narrative known as literature. A benefit of this approach is that it allows us to engage in a detailed, tangible way with perhaps the most fascinating aspect of counterfactual scenarios—their ability to spark the imagination. This most human and mysterious propensity evidently fascinated Richards, who touched on it in the *Principles*, and explored in depth in his later monograph on Coleridge ([1934]1960). More recently imagination has emerged as a cognitive process that is more accessible, and even more fundamental for our response to narrative than was previously realized.

So, what assumptions lay behind Richards' model? For a start, he confronts the elephant in the room early, proposing a level of identity between mental operations and the physical material of the brain. In *The Principles*, he brooks no argument on the issue: "That the mind is the nervous system, or rather a part of its activity, has long been evident" (75). On this, as so many other matters, he was an outlier, given the cultural and intellectual context of the 1920s. This position of "identity" between mind and brain, now a commonplace among brain scientists, is still sometimes contested amongst philosophers, and was then even more provocative.

Richards' use of an explicit graphical model of mental processes put him far ahead of the field. Despite his caveats, he was boldly imagining the events behind literary response at a level of sophistication only recently attained by psychology or neuroscience. So, where did this amazing model come from?

He acknowledged later in life that it was inspired by the ideas of a neuroscientist. Charles Scott Sherrington was awarded the Nobel Prize in 1932, along with Edgar Adrian, for "discoveries regarding the functions of neurons." In the course of a long and illustrous career, he coined the terms "synapse" (the functional connection between neurons) and "proprioception" (the brain's sense of the state and position of the body), and was recognized for his work on reflexes, defining the interaction of neuron pathways in these fundamental sensory to motor mechanisms.

In his 1906 book, *The Integrative Action of the Nervous System,* Sherrington described how information within and between reflex pathways combines to achieve properly coordinated behavior. He noted how, for example, if you accidentally step on a sharp object, you will flex the affected leg and withdraw that foot, yet unconsciously and almost instantaneously extend the other to maintain balance. To explain this, he demonstrated that these two actions are coordinated into one seamless reflex response by nerve cell circuits in the spinal cord.

Sherrington had been a student, then a fellow, at Cambridge before 1900. His legacy was deep, and would have been well known to Richards, who arrived there in 1912. Much later, in a provocative aside, Richards told his biographer that the model in *The Principles* came about because he "just translated Shelley into Sherrington" (Russo 1989: 179).[1] The profound imprint of Sherrington's neurophysiology is evident in the language Richards used to describe his "hieroglyph," when he outlined the competing "impulses" flowing through the system, and the resultant states of "equipoise," or balance, between them. This closely paralleled Sherrington's sense of the interactions between neural pathways when the rival signals underlying reflexes, positive (excitatory) and negative (inhibitory), are integrated to ensure coordinated control of posture and movement. Once again, Richards' model incorporated the latest neurobiological thinking of his time.

At this point, almost 100 years on, we can see that his early foray into neurophysiology foreshadows a number of recent discoveries about the cognitive processes behind reading; his diagram turns out to usefully depict what might occur when we respond to textual stimuli, even if the tools were not yet available to chart it in detail. Indeed, it is this very quality of condensation and suggestion that makes graphic illustrations, maps, and diagrams so widely used in science. Even when a map is conjectural, it can stimulate conceptual leaps and causal narratives.

It might be argued that some of the more far-sighted features of the "hieroglyph" are fortuitous, but they are no less remarkable for that. In an analysis of Richards' place in literary history, David West (2013) highlights the Darwinian assumptions underlying the *Principles,* which in turn open the way to pursuing narrative comprehension as an emergent cognitive process, which can also be set in an evolutionary context. In many ways, it also paves the way for those later critics who reject the notion of fictional narrative as a mere luxury, epiphenomenon, or "spandrel" but rather see it as conferring some selective advantage, allowing us to "develop our mental capacities, so that we are better adapted to our environment" (28). West goes so far as to label Richards a "protocognitivist," whose ideas would not surface again until the 1990s, pointing to his influence on cognitive literary criticism by noting that Peter Stockwell's contemporary work in cognitive stylistics "reads like a blending of Richards' *Principles of Literary Criticism* and *Practical Criticism*" (14).

Above all, it's evident that Richards looked forward to a more systematic analysis of the reader's experience. Indeed, there are those who are still using, and refining, some of his empirical methods (e.g., Goodblatt and Glicksohn 2010). He was not only an early proponent of a psychological attitude to literature, but the first to attempt an explicit model in which cognitive literary functions were articulated at a neurobiological level.

Considering all of the above, you could argue that neuroscience was actually "there at the birth" of modern literary theory. If Richards had been writing six decades later, it's not hard to see him collaborating with neurobiologists on fMRI (functional magnetic resonance imaging) scans during reading, or cognitive scientists researching memory.[2]

Richards' legacy is, in part, an early effort at thinking outside conceptual boundaries. However, the hard truth is that his early sense of the potential for cross-fertilization between disciplines failed to blossom with other critics of his generation, or even to bear fruit in his own later work. He made some references to his diagram in subsequent writings, but eventually it faded from active consideration, his own or others. Yet perhaps this is no surprise, given its boldness. Richards was far ahead of the field. Neuroscience itself would not be mature enough to elaborate on his model for more than half a century.

To understand the fate of his hieroglyph, we need to understand that Richards' model emerged in a period when the school of thought dominating psychology was behaviorism, which explicitly rejected analysis of what was going on inside the organism, or the brain, in favor of the study of overt behavior. Mental imagery and imagination lay well outside the range of the behaviorists' own theoretical maps. Here be dragons.

Behaviorism was to set psychology's agenda for almost five decades. In addition, the biological study of neurons—how they encode and transmit information electrically and chemically—could not make real progress until the 1940s and 1950s when technology provided the means to easily record electrical "messages and conversations" between nerve cells. Thus, any systematic or experimental approach to what was really going on inside the "black box" of the brain was not yet possible.

From the perspective of the arts too, the neurobiological aspect of Richards' work had little chance of being well received. To most, he had committed a cardinal error, even a category error, in pursuing an empirical methodology. Echoes of this attitude remain. Terry Eagleton (2002), reviewing Richards' contributions to the field, made the waspish aside that some of his work "smacks of the laboratory."[3]

Yet much has changed. The following chapters will look at interactions between narrative and the brain sciences at a number of levels: behavioral observations, systematic analyses, the role of pertinent brain regions, and, in some cases, even "drilling down" to what individual neurons are doing. In the course of our journey, we will show some of the innovative tools that have become available, together with the consequences for writers when literature is viewed from a neural perspective. Conversely, we will touch, albeit briefly, on what happens when science is seen from a narrative angle: a recurring theme and organizing principle of this book is that all such interactions are bidirectional. In chronological terms, our focus will be mainly on the twentieth century and beyond.

Cultural Context

The writings of Richards, and his openness toward science, were but a "moment" in a longer cultural discussion. As far back as the Enlightenment, there was a growing sense of a schism between the arts and sciences. An early alarm was sounded in 1709 by Giambattista Vico, and over time the acceptance of an unbridgeable divide gained acknowledgment, with some notable exceptions.[4]

The best-known recent airing of concern about the schism was *The Two Cultures* by C. P. Snow—a book based on his 1959 Rede Lecture (Snow [1959] 1998). The book has been so frequently discussed that it hardly warrants detailed consideration here, but the central thrust is that Snow decried what seemed to be a "gulf of mutual incomprehension" between literary and scientific thinkers. Unfortunately, his own attitude embodied the binary opposition he ostensibly deplored, by portraying scientists as forward-looking and socially conscious, while denouncing literary intellectuals as "natural luddites." One prominent critic who hit back was the Cambridge scholar F. R. Leavis (1962) who deplored Snow's overconfidence, but more substantively, his undeveloped understanding of what literature could achieve. The "debate" between Snow and Leavis stimulated a round of "culture wars" that would preoccupy Britain for many years.

To a remarkable degree, the Leavis-Snow controversy was a replay of one about eighty years earlier. In 1882, Matthew Arnold also delivered a Rede Lecture, entitled "Literature and Science." In it, he responded to T. H. Huxley, who championed the virtues of a scientific education, at a time when science in England had the connotation of a "vocational and slightly grubby activity" (according to Stefan Collini 1998: xiii). Arnold's view was more inclusive, conceding that, while the study of canonical works of science, such as *The Origin of Species*, was needed, so too was the study of great literature. These two cultural debates "bookend" Richards' early thinking in *The Principles* by about forty years on either side and his thinking at this time was closer to an Arnoldian view, recognizing that science was ascendant, but nonetheless positing an absolute need for "letters" and literature. Once again, critical debate would not catch up with this multidisciplinary stance until much later.

Even on the fiftieth anniversary of *The Two Cultures*, evidence of a continuing schism between humanism and science showed little sign of abating (Krauss 2009, Collini 2013). Stefan Collini, in his 1998 introduction to Snow's writing, in something of a stretch, suggested that a future resolution might come from theoretical physics. What neither Snow nor Collini foresaw was the rapid advance in cognitive psychology and neuroscience, and how they might advance our thinking about narrative literature, and our engagement with it, offering a bridge across domains. The evidence in chapters to come indicates that many critics are now acting upon Richards' early insights.[5]

The Neuroskeptical Turn

In 2005, the journal *Nature* published an issue with a special section on the recent engagement of artists with scientific material, and of scientists with the arts. Four articles were devoted to the topic of literature and neuroscience, a strong indicator that "the second culture" of the humanities, as Snow referred to it, was neither uninformed nor uninterested in neuroscience.

Each of the reports on literature featured a prominent writer grappling with scientific ideas, A. S. Byatt being one. She described her education at Cambridge where, shortly after she graduated, the two cultures dispute erupted. As she put it, "that was a university-political battle. Leavis held a semi-religious belief that the study of English literature was at the centre of the idea of university. As a young woman I felt that it was only possible to study literature and try to write it if one had no such exclusive semi-religious attitude" (Byatt 2005: 294).

It wasn't that Byatt sided with Snow in the controversy, but that she wished to supersede it altogether. Writers she cared about—for example, George Eliot, Samuel Taylor Coleridge, and T. S. Eliot—were themselves fascinated by the scientific work of their time, she pointed out, revealing her own increasing immersion in the works of thinkers like Richard Gregory, Jean Piaget, Noam Chomsky, and Jean-Pierre Changeux. Fundamentally, she said, what excited her were descriptions from both literature and science of what the *experience* of cognition is like, taking many forms, from metaphors to explicit graphical models.

After the piece in *Nature*, Byatt published an essay in the *Times Literary Supplement* elaborating further and articulating how, for her, "[John Donne's] great love poems stir both *body and mind* in an electric way that resembles nothing else" (Byatt 2006: 247, our italics). Byatt is a particularly conspicuous example of a writer who has found the sciences of brain and mind to be both a source of inspiration and a means of understanding. In her own writing, fictional and discursive, she demonstrates both inspirational and critical uses of brain science, together with a keen interest in embodied thought, hence her critical focus on writers who convey "the feel of thought."

In the *Times* essay, Byatt also expressed her dismay that psycholinguists seemed uninterested in the mental impact of "words" at any fundamental level. By contrast, when she turned to the work of neuroscientist Jean-Pierre Changeux, she discovered incisive analyses on topics such as perception, the construction of mental objects, and the formation of memories, outlined in terms of neuronal connections. As she said:

> It has been much easier [for me] to think about this aspect of poetry since there have been increasing quantities of information available about the myriads of neurons in different areas that fire when things are perceived, are reinforced by connections with previous perceptions and previous connections, and make up the constantly changing matter of mind. (249)

Throughout, she scrupulously labels her own neural speculations as just that, awaiting corroborative research.

In response to Byatt's exploratory essay, self-styled neuroskeptic Raymond Tallis (2008) issued a sharp rebuke: "Bellowing in a rage when one discovers that the toilet paper has run out, and someone has neglected to replace it, would involve the very same processes Byatt invokes to explain the particular impact of the poems of a genius, if such processes do occur" (14).[6]

Such fulmination is an extreme version of neuroskepticism: simply defined as a distrust of neuroscientific claims about any but the most technical of questions, along with a visceral rejection of any scientific enthusiasm from humanists. Some degree of neuroskepticism is unsurprising, even warranted given the inordinate amount of sensationalist and simplified press attention brain research has received in the past decades. Moreover, there are certainly grounds to be wary of some uses of functional neuroimaging, relating to over-interpretation, premature or biased reporting, inappropriate applications, and inflated claims of normative significance (Rachul and Zarzeczny 2012, Satel and Lillienfeld 2013).

To most voices of neuroskepticism, one can simply say—fair enough—at least as a corrective. Science itself is, after all, a process of deploying systematic skepticism. And where there are neuroskeptics, there are also "neuroadvocates": those who push a science agenda too far, assuming literary culture needs to be "saved," and that only neuroscience can do it. The key point is that debate around the significance and bearing of brain science is not new, and is unlikely to disappear.

Inspirational Reach

Despite the wrath of skeptics, many novelists have been receptive to brain-mapping—are indeed intrigued by the findings of brain science as a whole—and this has been reflected, formally and thematically, in their work, even when such findings seem to impinge on territory that was formerly their exclusive preserve. They have sought to adapt by assimilating, spinning narratives even from their own fears. What, they ask, can storytelling reveal about our inner selves in the age of brain scans, neurotransmitters, and cognitive architecture? What is the role of introspection and imagination? Is neuroscience now adding value to our understanding of literature, or is it the other way around, literature adding value to brain science?

Certainly, this latest breakout of anxiety from the humanist camp— think of Arnold then Leavis—has had a negligible effect on the flood of literary narratives on brain-related ideas and pathologies. Novelists who have delved into cognitive and neuroscientific themes include Rebecca Goldstein, Mark Haddon, Siri Hustved, Nicole Krauss, Jonathan Lethem, Simon Mawer, Richard Powers, Will Self, and Edward St Aubyn to name a few. For the most part, these writers (and others, see Ortega and Vidal 2013) engage with the debate either by creating a character with a neurological or psychological condition that acts as a portal into current disputes—Baxter in McEwan's *Saturday*, Andrew in Doctorow's *Andrew's Brain*—or by crafting a plot which forces the reader to often uncomfortably engage with topics such as determinism, the mind/brain dichotomy, or the role of storytelling itself. Again, none of these are new preoccupations (see Anne Stiles 2007) but in the light of developing scientific technologies and some compelling findings, they have been revivified in fresh, sometimes disturbing ways.

The reaction of novelists to "the other culture" across the arts/science divide is by no means restricted to the use of scientific protagonists, the incorporation of scientific plots, themes, or even nightmares (cf. Mary Shelley and the recent surge in post-apocalyptic fiction). It can also be a matter of stance and disposition. In 1876, George Eliot wrote to Joseph Frank Payne that her novels were simply a set of experiments in life, and this quasi-scientific outlook is one many other writers have identified with over the years. Byatt has repeatedly referenced Eliot's work as exemplary in this respect.

Interactions between the two camps go well beyond what was traditionally called "inspiration," however, and occur both ways. Psychologist and novelist Keith Oatley looks at fictional narrative primarily from an instrumental perspective, seeing it as something analogous to a "flight simulator" where a reader can enter convincing scenarios in a virtualized context. This is the novel as research-lab and hypothetical paradigm, with the reader as sometimes unwitting subject. Inventive and exciting though such analogies are, the danger is that when literature is regarded in a utilitarian light it can appear more like a dummy-run for life, an offshoot of evolutionary adaption, and jeopardize its intrinsic value. In such a reading, plots can become little more than a series of pragmatic hypotheses, or "trials" in the medical sense, serving empirical goals. The novel surrenders its artistic autonomy.

It is hard to overstate the degree of controversy surrounding not just the methods, but even desirability of applying a scientific mindset to traditionally literary pursuits. Critics, philosophers, and practitioners of both narrative and science rightly feel there is much at stake.

Point Counter Point

For more than two decades, E. O. Wilson, an eminent biologist and one of the leading experts on social insects and biodiversity, has been writing on the desirability of unifying the sciences and the humanities. He relates this to what he calls the Ionian Enchantment: "a belief in the unity of the sciences—a conviction, far deeper than a mere working proposition, that the world is orderly and can be explained by a small number of natural laws" (Wilson 1998: 4).

It is undeniable that the natural sciences form a highly successful, concatenated system for observing the world around us. Biology addresses living systems underpinned by chemistry, and chemistry in turn relies on the laws of physics. These disciplines can be seen as levels of description that cannot be reduced, one to the other, without loss—so multidisciplinarity across these sciences makes sense.

Given the success of the scientific method, Wilson proposes the consolidation not only of the sciences, but of all knowledge by extension of the methods of the natural sciences to the humanities. He applies the term *consilience* to this grand project. Consilience is an example of interdisciplinary integration writ large, a significant feature of which is methodological exclusivity.

For Stephen Jay Gould, this contains the assumption

that the chain of reductionism heretofore so successful in stretching from particle physics well into the reaches of biological complexity [is] poised to make its boldest move upward—starting with ... initial successes in beginning to understand the workings of the human brain, and then moving through the social sciences and eventually, and ultimately, into the traditional humanities of arts, ethics and even parts of religion. (Gould 2003: 16)

By reductionism, Gould is referring to the strategy implied above of explaining complex phenomena by analysis of simpler components: organs explained by cells, cells by molecules, molecules by atoms, etc.

Wilson's ideas on consilience between the sciences and humanities have been around for more than two decades. While often referenced, they have not gained quite as much traction as Wilson clearly hoped. In literary studies it has been generally ignored, with one notable exception: Darwinian literary criticism, as conceived by Joseph Carroll and advanced by Jonathan Gottschall and others. In Nancy Easterlin's summary of this development (2012), she identifies one reason this might be so, pointing out that Wilson's overarching theory fails to offer a comprehensive basis even for a biocultural literary approach because it largely ignores the aim of literary criticism, and narratology, which seeks to analyze human experience through the range, style, and emotional depth of stories: a venture based not on atoms, but on subjective interiority. Wilson's approach largely neglects the "how" and "why" of literary criticism, and the role of value judgment, of irreducible subjectivity, at its heart. This is not to say that the *goal* of consilience lacks appeal. Ian McEwan, shortly after completing his novel *Enduring Love* in 1998, revealed that "my own particular intellectual hero is E. O. Wilson. He's a biologist. He wrote 'The Diversity of Life,' and that was just genius. The thing that really interested me was the extent to which scientists are now trespassing into other areas" (interview March 31, 1998, Salon.com). Yet by 2009, his wholesale enthusiasm is tempered by a more judicious sense of traditions working in tandem:

> As a novelist I suppose that one of my central concerns is the investigation of human nature, and the biological materialism of Darwin fascinates me because it's opened up so much in the way of explanation. ... that thinking biologically as well as [of] ourselves as cultural products is central to both one's curiosity about who we are and curiosity about how our science is going to unfold in future years. (Ian McEwan on Darwin, *The Science Show*. ABC, August 1, 2009; transcript March 16, 2010)

The general lack of uptake for consilience in the humanities is partly a matter not only of substance, but also of tone. Although Wilson praises the goals of humanistic thought, he seems unappreciative of its accomplishments and methods. So what does consilience propose? Ultimately, a solution where humanistic inquiry becomes assimilated into the scientific domain: not so much an accord, but an annexation.

Two examples demonstrate the circumscribed impact of consilience. Like writers, visual artists have been increasingly engaging with science and scientists—from a position of equality and parity of esteem—not as any kind of "poor relation." The section in *Nature* described earlier hints at this. A hallmark of recent art-science interactions is the sharing of supposedly contested, but often complementary, topics and approaches, even resulting in shared laboratories, galleries, and performance venues. Funding agencies, recognizing this growing détente, have been actively supporting projects that encourage and enhance work at the interface. So, it might be said there is some "consilience" here, but with a very large dose of mutual respect *and* celebration of methodological diversity (e.g., Swanson et al. 2008, Conway 2012).

One field above most others provides an object lesson in the need for comprehensive ("binocular") vision, devoid of full-blown consilience. Donald Peterson (2004) has described how, at the outset of modern psychology, William James "understood the folly of hyperskeptical extremes" (198). When the study of human minds emerged as a science it also developed as a healing practice with deep humanistic roots and, within itself, psychology has managed to sustain these two complementary cultures. The reductionist, statistical approach of research labs and psychopharmacology coexists alongside case-based, narrative, and therapeutic approaches. Within applied psychology, success in treating mental disorders tends to produce respect for reductionism. At the same time, clinical realities leave most psychologists with a healthy sense of a science with limits. As Peterson noted, "we cannot control the systems which we examine—the lives of human beings—and would not want to if we could" (201). Certainly, there have been frictions, amid calls for a more rational, biologically based way to categorize and diagnose psychological conditions, but there is also a pressing need to deliver clinical care with a human touch and a sense of how individuals relate their stories. This is an internal tension that psychology lives with, and even benefits from, without any apparent need for consilience.

Despite the successes of reductionism within the natural sciences, its application to complex systems is not without difficulties. An excellent example comes from one of the most fundamental concepts in brain science: the reflex. Sherrington suggested a reflex was like an "atom" of behavior (or what cognitive science calls a primitive): a simple unit from which more complex units of behavior are composed. However, after years of study, he ultimately realized that while reflexes can explain some simple behaviors—especially those controlled by the spinal cord and brainstem—these "atoms" will not explain higher cognitive functions: helpful at one level of analysis, they prove inadequate at others. We see this failure of reductionism at other levels of analysis. When memory is considered as a fixed record like a photograph or video clip, the analogy quickly unravels. As you will see, this technology-inspired, reductionist model is just too simplistic to convey the reality of our varied and volatile memory systems, and misses their most intriguing attributes (Chapter 9).

So, how might living (thinking, feeling) systems frustrate attempts at explanation using *nothing but* biology and chemistry? This question marks one important starting point for developing a mature cognitive science, as stated by pioneers in the field: "Just as biochemistry is not 'simply' physics ... thinking is not 'simply' neurophysiology but calls for levels of theory that can be linked to neurophysiology only through a sequence of connecting theories most of which have not yet been fashioned" (Simon and Kaplan 1993: 6). In short, higher-level cognitive theories are necessary for explaining our mental life, and the central role played by narrative. Nonetheless they won't be relevant if they fail to connect at all with neurophysiology.

There are other reasons why we need higher-level theories. Stephen Jay Gould, for one, pointed to two axioms that must be taken on board before dealing with brains. First, as systems become more complex, they often demonstrate *emergent properties*—novel phenomena not strictly predictable from reductionist analyses at a lower level. These properties derive from systemwide interactions, especially those that are non-linear or non-additive. A good example would be broad affective states, including typical emotions. Despite the fact that we unquestioningly experience states we call "fear" or "empathy," when we look (and scientists have) there are no fear neurons or empathy neurons to be found. The "grail" does not exist. There is no "fear center" in the brain, rather a number of regions participate in generating emotion from distributed activity across networks, requiring new levels of theory (Chapter 7).

Second, biological systems demonstrate *contingency*: some of their features are explicable only by recourse to the past. This observation, that the course of evolution is constrained by history, is visualized by Stephen Jay Gould in the thought experiment of rewinding the "tape" of evolution backward and replaying it. So, for example, the emergence of aquatic organisms onto land did not ensure the rise of primates or even of mammals—it could easily have been otherwise. Contingency is relatively unimportant in physics and chemistry, but an essential feature of biology and therefore a vital ingredient in the epistemology of the brain sciences. As Ernst Mayr so clearly pointed out, such contingency—historicity—is what makes biology so distinctive from the physical sciences (interview with Mayr: Lewin 1982)—and its implications resonate through every study of brain and mind.

Approaching Scientific Imagination

On the question of influence, it would be a mistake to think this has all been one-way, from science to the arts. Our human need for narrative structures and techniques has had a profound bearing on how science—perhaps especially neuroscience—is proposed, explicated, and expressed. One of the most obvious ways that literature has influenced neuroscience is through the subtlety of its descriptions of human perception, social conduct, and mental disposition. As James Wood put it (2008, 2015), the most enduring narratives are always based on a detailed knowledge, understanding, and evocation of behavior. Such an "empirical slant" in fiction Wood sees as characteristic of literary realism, a style distinguished by what he has called "serious noticing," citing writers like Chekhov, Tolstoy, Joyce, Henry Green, Saul Bellow, and Marilynne Robinson.

Whether "realistic" or not—and it's a notoriously slippery term— few would argue against the idea that literature, through the rendering of consciousness, sensation, and feeling, can provide imaginative access to subjective states that are not (and may never be) directly accessible to neuroscience. As computer scientist David Gelernter put it, "Keats or Rilke, Wordsworth, Tolstoy, Jane Austen [as] 'subjective humanists' can tell us, far more accurately than any scientist, what things are like inside the sealed room of the mind" (2014: 22).

Even within science, subjective, literary, and imaginative influences can be clearly discerned when its processes are viewed realistically rather than idealistically. Two scientists who have reported insightfully on these matters are biologist Robert Root-Bernstein and chemist Roald Hoffmann.

For example, biochemist and artist Root-Bernstein has said, "The road to objectivity in science is paved with subjectivity ... [yet] the object of experiment, proof and analysis is to expunge this subjective residue from the final statements of scientific fact" (1997: 6). He identifies an early and necessary imaginative phase in scientific thinking that operates "pre-logically" through images (visual, kinesthetic) and via emotion, which is only later translated into words for communication purposes and for tests and conclusions. This second logical and linguistic phase has led us to ignore its genesis in imagery and emotion. Science often chooses to ignore its roots.

Nobel laureate chemist Roald Hoffmann (2012) has gone further to suggest that these primary creative stages in scientific endeavor depend upon on the most basic act of literary imagination, forging metaphors, which permeate

> the daily practice of doing [scientific] research—in the way scientists generate hypotheses, theories and experiments. But ... people don't much admit to it. My observation is that scientists sanitize their papers to remove as many explicit admissions as possible of the fecund generative utility of such metaphors. Why? Because metaphors are (mistakenly) thought to impress no one—they are not mathematicizable; they are "less rational." (249)

By his account, even in the act of translating the pre-logical into the logical, the construction and articulation of rigorous tests and papers, scientists cannot avoid the literary aspects of our thought: especially the need for compelling narrative. This is an area that has massive implications and is gaining consideration in its own right, but the writings of Hoffman and Root-Bernstein provide an accessible entry point.

Assuming the influence of narrative imagination on science, it makes sense to go back and first address how it is that narratives create meaning and to outline some of the features common to all storytelling. In David Herman's *Storytelling and the Sciences of Mind* (2013), he begins by acknowledging the centrality of narrative in human history and suggests some of the reasons for its universal emergence. As he sees it, "Narrative—from before the start of literate culture—has served as a support for the formulation, systemization and transmission of communal as well as personal experiences and values" (23). In a comprehensive analysis, he enumerates specific functions that narrative performs, including: the "chunking" of experience into psychologically manageable units, related to the Aristotelian divisions of "beginning," "middle," and "end"; the provision of a virtualized realm in which to grasp causal relations; the opportunity for social problem-solving by way of parallels to "real-life" scenarios; the dissemination of knowledge and folk memory; and the creation of an arena in which to practice the ascription of intentionality and state of mind to others (via what is usually called "mentalizing" or "mindreading"). Narrative, in all its forms, has proved durable, portable, and adaptable, and there may be a plethora of neural and cognitive reasons why this is so. Support for many of Herman's points is found in anthropological data as described in Chapter 1. As others have suggested, we live storied lives.

Approaching Narrative Imagination

The term "narrative imagination" can initially be understood in two ways: first, the creative process by which writers explore ideas, emotional states, and their meanings through stories, drama, or verse; second, the process of assimilating such meanings, and constructing inner "worlds" via reading or listening. The second is squarely centered on reception, or "reader-response," a primary consideration of I. A. Richards.

Indeed, in *The Principles* Richards devotes much space to the theme of imagination (it is one of the longest chapters in the book—twice the length of his compact, forward-looking discussion of memory). He notes six common uses of the term: (1) The production of vivid images, perhaps its core meaning, extending back at least to Aristotle; (2) The use of figurative language; (3) The sympathetic reproduction of other people's states of mind: a lovely anticipation of recent interest in "mentalizing"; (4) Basic inventiveness; (5) The connection of things ordinarily thought of as disparate, perfectly exemplified by scientific hypothesizing; (6) Finally, imagination that is distinctively poetic. Shortly after publication of *The Principles*, Bertrand Russell highlighted the importance of inventiveness in this breakdown, and its relation to the allied power of memory: "The essence of imagination ... does not lie in images ... the essence of imagination, I should say, is the absence of belief together with a novel combination of known elements" ([1927] 1960: 199).

A more recent voice from philosophy, Paul Ricoeur, has—with an eye toward literature—teased out elements of the mental actions we ordinarily lump together as imagination (Taylor 2006). First, he pointed out that the idea of a mental image as a copy of perception is a *reproductive* form of imagination, where the real interest lies in *productive* imagination. Under this rubric he places *poetic imagination*, but also *social and cultural imagination* (e.g., Orwell's *1984,* Atwood's *Handmaid's Tale,* and other dystopian or utopian visions), and *epistemological imagination*, creativity that seems to extend beyond the known or real, both the factual and the counterfactual, as seen in fiction, and some scientific models. Richards, Russell, and Ricoeur all share a belief that imagination is essentially constructive. In this sense, it shares characteristics with memory, which, it turns out, is also a constructive process. Given these precursors, it is hardly surprising, then, that evidence from cognitive psychology and neuroscience suggests our engagement with narrative—the creation of our storyworlds— is constructive, too (Chapter 6).

Moving Forward

Richards' writing and his early hieroglyph point to two very different ways in which scientific models may have bearing upon the narrative imagination; we would call these an inspirational mode and an instrumental mode. In the first, scientific views and methods make an impression on the shape, content, and even narration of a literary work, informing its thematic preoccupations, its plot, characters, setting, chronology, and style. The whole genre of science fiction is a testament to the inspirational mode. In the realm of literary fiction, there is an embarrassment of riches when it comes to examples of novelists who have turned to science, often brain science and its latest controversies, for ideas: the likes of Jules Verne, H. G. Wells, Karel Capek (who invented the word "robot"), and Ursula K. Le Guin spring to mind.

However, science can also be useful when considering the audience or reader. In this instrumental usage, scientific models can provide an evaluative and analytical tool to help us grasp how a narrative works upon us: how it is comprehended, commands attention, realm, and triggers empathy. These last two phenomena have been much studied recently and may be related. Transport, the process of losing awareness of our immediate surroundings as we become absorbed in a story, is of interest to not only psychologists but everyone who reads (see Nell 1988, Gerrig 1993 for fine examples of empirically grounded studies). At its best, transport has us lifted from our current location and "moved" to somewhere else, by way of another consciousness. We are suddenly "in" Hogwarts, or Henry James' Boston. Recent experiments suggest that transport may facilitate empathic responses (Chapter 7). Beyond this, without its ability to "teleport" us across space and time, it is hard to see how literature could have maintained its high cultural value.

Yet ever-finer understanding of the brain's involvement will never negate the magic of reading, or entail any dismissal of literary criticism, which will continue to resist subsumption into scientific methodology, despite the best efforts of Wilson and others. Nancy Easterlin's corrective is timely: "The very nature and diversity of literary artifacts, which are themselves only fully constituted via a complex cognitive process of production and consumption, a process inherently interpretative, militates against a programmatically scientific approach to literature" (Easterlin 2012: 6). The approach in this book is broadly in keeping with Easterlin's biocultural stance, combining traditional humanist methods with cognitive neuroscience and evolutionary social science as they prove useful, revealing, and relevant.

There is a distinction here which will assist the journey ahead: between a *multidisciplinary* approach, which is open to the juxtaposition and deployment of information and methodologies from various fields and an *interdisciplinary* one, which seeks some level of integration, often a deep one, between disciplines. The latter has sometimes masked a drive for hierarchical realignment across fields, involving the superimposition of inductive criteria, such as falsifiability, upon humanistic pursuits. Such a push would ultimately invalidate the subjective and deeply personal elements of imaginative engagement with literature, and is indeed what neuroskeptics most fear (see Bernini and Woods 2014). While we argue that consideration of the brain sciences is a valuable, even necessary, adjunct to humanist criticism, it is surely not sufficient in itself.

Returning to this distinction between the "inspirational" and the "instrumental," Richards' model—his extraordinary "hieroglyph"—clearly falls into the latter category, bringing a scientific awareness to bear on narrative reception. In this sense, his sketch was a multidisciplinary map of possibilities. He passionately believed it was a glaring omission to ignore the workings of the brain in the process of analyzing the creative mind. His hieroglyph is also a landmark: one of the first suggestions that neuroscience could provide a different sort of understanding of literature. But pioneers are often disregarded and disbelieved, and the fruits of his bold thought-experiment were to wither on the vine. Even today, many critics feel discomfort faced with such a materialist stance: the charge of "scientism"—of a naive, almost mystical trust in the applicability of science—is never far away.

We know even from basic anatomy that the only evidence of the external world that enters the vault of the skull is in the form of signals in our nerves, and there is always a gap not only in kind, but also in time, between objects in the world and our mental representations of them. Our maps are permanently out of date, our percepts hopelessly late for consciousness, in Ian McEwan's phrase. Yet narratives—oral and written—seem to offer a way out of this dilemma, laying a causal web over the fictional realm, offering a comprehensibility only rarely glimpsed in real life. At another level, too, the story exists in an eternal present, awaiting only the reader or listener to bring it to life. Magic indeed. The fact is neuroscience does not yet and may never have a better way of describing what it is like to be someone than narrative. One function of the "neurocognitive turn" in critical thought over the last three decades has been to see how far science can overcome its own limitations (Spolsky 1993).

Looking from the humanist side, Mark Turner (1991) exposed what he saw, then, as a gaping hole at the heart of the critical endeavor:

> The triumph of literary criticism has given us a concept of literature as the product of circumstances—biographical, psychological, political, economic and social—not as a product of the capacities of the human mind. We do not ask what is a human mind that it can create and understand a text? What is a text that it can be created and understood by the human mind? (16)

These words could just about have been penned by Richards almost seventy years before when he predicted that an understanding of brain science and neurophysiology could deepen and enrich our appreciation of what literature does, and how it does it.

Taking our lead from him, we should return to some basic questions. What is our current understanding of the value of narratives? Where have they even come from? How should we regard the role of mental imagery in our reception of narrative, and how, if at all, do these conjured images relate to our primary acts of perception? We also need to ask how far and how deeply narrative is "embedded" in our history and psyche, given our niche as a highly social primate? Such questions will set the stage for a deeper look at brain, cognition, and narrative in the sections ahead. But certainly, we will move beyond the oppositional binary of two cultures, and we must add: beyond consilience. Let's begin to explore how that might be done.

Notes

1   While Sherrington is best known for studies of the spinal cord, he also worked on the organization of mammalian cerebral cortex and clearly knew of its layered organization, reflected in Richards' model. Sherrington also showed that spinal integration is modulated by upper level control from the cerebrum.

2   Gregory Jones (1993) pointed out that, amazingly, in chapter 14 of the *Principles*, Richards described complex spaces that could represent memories. These were basically "attractor networks" that would not be discovered until late in the twentieth century, based on early connectionist modeling at mid-century.

3   However, Eagleton's review of *I. A. Richards Selected Works, 1919–1938* is, on balance, almost positive, perhaps indicating Richards will be appreciated anew in this century.

4   Cognitive historicist studies have documented significant inspiration/influence of neurological thinking on writers from the romantic period, such as Coleridge. See the very clear analyses by Alan Richardson (2001, 2011).

5   Easterlin (2012) has produced an interesting overview of the two cultures
    debacle that links it to larger trends in the development of literary studies.
6   Tallis (2008: 14). Recent, heated discussions around neuroscience and
    literature include suggestions that those who consider brain research to provide
    insights into literature are *neuromaniacs*, the neuroscientific study of human
    behavior is *neurobollocks*, and humanists should defend us against the *Fausts
    of Neuroscience* (Noe 2011, Poole 2012, and Walden 2012, respectively). Also
    see Tallis (2011).

PART I

Motivations to Explore Our Storied Mind

Humans have deep connections with stories and the narrative arts. Psychology and cognitive science have already made great contributions to literary analysis, and neuroscience is starting to contribute significantly to the understanding of our storied minds. Other fields also have much to contribute, and gains are being made from a multidisciplinary approach to story, its variety of forms, and its effects on us. Perhaps the greatest return has been a better grasp of the reciprocal dynamic by which culture molds, and is in turn molded by our brains and minds, and how the imagination is an important mediator of this.

1

The Scheherazade Syndrome

In 1994, A. S. Byatt published a novella, *The Djinn in the Nightingale's Eye,* about a contemporary woman with a professional interest in stories. The protagonist, Gillian Perholt, is first seen on a plane leaving London for Ankara to attend a meeting of narratologists. As the plane ascends through the clouds, the novella's own narrator lays out Gillian's backstory and the nature of the tale of which she is part. "Her business was story telling, but she was no ingenious queen in fear of the shroud brought in with the dawn" (95). The reference is of course to Scheherazade, narrator of *One Thousand and One Nights*, the centuries-old collection of folk tales from the Middle East. The allusion also points to the temporal and thematic scope of the novella to come. Behind this is a presumption. For it is taken as given that Byatt, Gillian, and indeed the reader share the propensity to be entranced by entirely invented characters and the experiential journeys they undertake, whether contemporary or ancient.

The story displays some playfully recursive maps within maps; a key reframing device being that, in the course of Byatt's story, Gillian in effect "enters" the very stories she is studying, mostly folk and fairy tales; the Djinn of the title turning up as a genie with whom she has a series of ever more puzzling and intimate interactions, and who, in the traditional manner, grants her three wishes, but also teaches her how to avoid the traditional folk-tale irony of regretting her choice.

Along the way we are given Gillian's scholarly take on narrative. We are told that she "was accustomed to say, in lectures, that it was possible that the human need to tell tales about things that were unreal, originated in dreams, and that memory had certain things also in common with dreams; it re-arranged, it made clear, simple narratives, certainly it invented as well as recalling" (203). For the Djinn is not only a dream-like element in the story, the Djinn *is* story, providing wish-fulfillment even as it offers enlightenment. We too may be puzzled by the intimate power of narratives, especially fictional ones, but we can no more ignore their connection to our mental lives, conscious and subconscious, than we can refuse to partake of three wishes.

Some of the most fundamental questions about narrative, and its effect on us, can be lost if we focus only on completed fictional works, their themes and structure, antecedents, and impacts. If a fundamental aim is to learn how sentences and paragraphs have impact, how they come to appear meaningful in the first place, or prompt the imagination, it is important to consider approaches where the focus is narrower. We need a map on a finer scale. The earliest European narratives, like the *Odyssey,* and the earliest novels, like *Don Quixote,* are journeys, a tradition that continues through the years in works like Conrad's *Heart of Darkness* and Cormac McCarthy's *The Road.* A journey is composed of steps. For a reader, making their own journey into a text, those steps are words. Narratives, of course, need not be word-based at all: as silent film, mime, dance, and graphic novels all attest.

Most of our attention will be given to written texts, often literary, always remembering that oral narratives far predate the invention of writing. We will also bear in mind Wolfgang Iser's view of literature as a cultural phenomenon occupying a unique intersection, "the interplay between the fictive, the real and the imaginary" (Iser 1993: 2). Yet fiction, besides pointing to "the real," has demonstrable effects upon your brain in the very act of conjuring up "the imaginary." So to provide the fullest account of a complex process, it is necessary to work with a range of survey tools (from narratological analyses and questionnaires, to brain scans and microscopy). Reading fiction is not only a temporal event, as we take in words, then lines, resolving them into scenes and episodes, but one that takes place at an emotional, rememorative, conceptual, and neurophysiological level.

For this reason, in the course of our investigation, after we've sketched some fundamentals about the choices available to authors, as they guide us into their invented worlds, we spend some time surveying the terrain of the brain sciences—sometimes even the physical topography of the brain—to supplement our understanding of narrative imagination from this perspective, which is after all fairly recent: outlining not only where we're at, currently, but where we're coming from and where we might be going, as readers and writers; concluding by looking at novelists who have begun to incorporate the findings of cognitive and neuro-science into their own work, in equal measure inspired and awed by the consequences for ideas of free will, consciousness, creativity, and the self.

A multi-level focus on reading is natural here: before we have the experience of a novel, a chapter, a theme, or even an image, we have eyes scanning a page, the remarkable, serial process of reading, and the highly selective stimulation of a vast multimodal "vault" of memory and association in the reader.

The use of complementary approaches, and the book's later return to discussion of metaphors, characters, and influences, is also an explicit recognition that traditional critiques of stories and literature are indispensable. Close-reading and exegesis not only retain a role, but are enhanced by the science. Hence they form part of the later chapters, shifting the frame, scale, and projection, upward and outward. Sometimes it is most helpful to take such a conceptual view, at others detailed empirical analysis is more appropriate. We need to analyze the soil and the rocks before we can begin to understand the forests and the towering peaks.

So, for the most fundamental questions about how stories "work" in a basic sense, we will shift between levels, often relating narrative structures back to the primary act of perusing a text, the comprehension of words, and the neurobiology of doing so.

In some respects, the critical stance here reflects Elaine Auyoung's analysis of the reading mind, *When Fiction Feels Real* (2018) where she posits "a new form of critical attention ... [that not only] approaches the words of a literary text ... as bearers of interpretive meaning, but as cues that prompt readers to retrieve their existing embodied knowledge, to rely on their social intelligence, and to exercise their capacities for learning" (18). Hence, this book often touches on reader response, how we assimilate the sensory and affective properties of "storyworlds," before shifting to more explicit, empirical considerations of reader's minds through brain organization and function.

Before we can consult these various maps, it is worth enquiring whether the journey is worth the effort—taking time to consider the significance of story, and narrative more generally in our lives, both as art and as everyday anecdote. As Porter Abbott (2008) has pointed out, our tendency is to think of narrative only in the former, more aggrandized sense, but narrative is the mundane currency of everyday exchange, the bedrock of social relationships, the salve of our pains, and the heart of our humor. It is common to almost all human interactions. As Jim Crace—a novelist considered later—puts it:

> All writers are doing is doing something formally, between hard covers, that all of us do informally as a necessary function of being human beings. We keep on imagining the future and reinventing the past ... it's a necessary function of human consciousness, to be a narrative person. (Interview, Crace 2012: 1)

A logical place for us to start looking at this "necessary function" is by establishing the natural habitat of narratives, beginning with oral storytelling, since that is where it all began.

The Ecology of Story

Recent observations from anthropology begin to explain what it might mean to say that stories are natural phenomena. Observation of storytelling in pre-literate societies may give some idea of humankind's deep history as a highly social primate: the relationship between talking and fire. Polly Wiessner

(2014) gathered information on conversations and their contents, among Ju/hoansi (!Kung) bushmen on the border between Botswana and Namibia. They are a small-scale society that has subsisted largely on foraging: a way of life similar to that of human cultures throughout much of our evolution, and preceding the cultural invention of writing about 5,000 years ago.

Humankind extended the daily time available for social interaction with the use of fire (something we likely have been doing for 400,000 years or more), and the question behind this study was elegantly simple: are fireside hours being used as a mere extension of daytime? From 174 episodes collected over months, the answer became clear. Socializing at night was not "more of the same," it was adding something new. Figure 1.1 shows the breakdown of conversation topics during daylight, and at night around the communal fire.

Wiessner found that day conversation is dominated by the practicalities of life: economics, complaints, land-rights, etc.; stories are only 6 percent, leavened with some joking. However at night there was a dramatic change: by far the most common verbal activity was storytelling, and fixation on the daily business of survival fell away. As Wiessner indicates, "Firelit stories centered on conversations that evoked the imagination, helped people remember others in their external networks, healed rifts of the day, and conveyed information about cultural institutions" (14027). Both men and women told stories, and Wiessner noted that this often included older individuals who seemed skilled in the art.

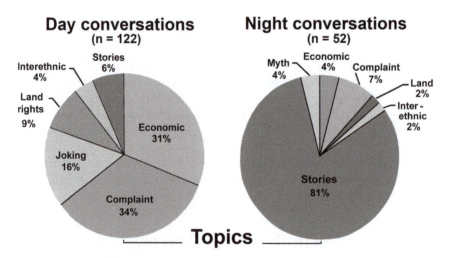

FIGURE 1.1 *The ecology of stories. The topics of day and night conversations compared. These data come from Ju/'hoansi (!Kung) bushmen of southern Africa (Botswana and Namibia). From: Wiessner (2014) Embers of society: Firelight talk among the Ju/'hoansi bushmen.* Proceedings of the National Academy of Sciences. *111:14027–35, Figure 1, used with permission of the journal.*

She is clear that stories not only entertained but also informed. The quirks and traits of tribe members, such as kin from prior generations, were refreshed and embellished, as were relationships. "Night conversations used multimodal communication with gestures, imitation, sound effects, or bursts of song that brought the characters right to the hearth and into the hearts of listeners" (14030). Furthermore, they typically were "about known people and amusing, exciting, or endearing escapades. Storytellers did not praise heroes or moralize; advancing oneself in the moral hierarchy or demoting others was avoided, as was any form of self-promotion" (14029).

When an anecdote was over, others in the group rehashed details, further embellished, and discussed it. The language of stories tended to be rhythmic, complex, and symbolic, with individuals repeating the last words of phrases or adding affirmations. Frequently, listeners "were stunned with suspense, nearly in tears, or rolling with laughter; they arrived on a similar emotional wavelength as moods were altered" (14029).

These results provide valuable pointers to the issues we will address. For example, Wiessner identifies some key attributes of storytelling: the visible suspense of the listeners; *memories* brought to life, especially in terms of kin relationships; *emotions*, moods, and *imagination* evoked; and behind this, often an implicit "moral" or lesson. Thus storytelling (and as we will see, listening and reading) touches on deep cognitive processes as well as having a social rationale. In short, Wiessner's study demonstrates the place of oral storytelling in our mental ecology (see also Chapter 3 and Figure 3.1) and suggests its important place in cultural and even biological evolution. Other critics, analyzing casual conversations in so-called developed cultures, have come to similar conclusions: "We spend 40 percent of our conversational time in spontaneous narratives" (Slade 1997: 7—and see also Boyd 2009, on the evolution of stories from play).

The central role of storytelling in the evolution of the human mind has not escaped literary treatment. It is impossible to read the intriguing report by Wiessner and not think of the kind of early human storyteller described in the novel *Gift of Stones* (see Chapter 2) who almost literally deploys his storytelling prowess as a "prosthetic" after he loses his arm, and hence ensures his practical usefulness in a flint-working tribe.

Wiessner's field-based study reminds us that if neuroscience is to play a part in the understanding of stories and literature, it needs input from other fields. This means taking anthropology, psychology, linguistics, and other human sciences seriously, and also attending to the theoretical insights of philosophy. Such a wide-ranging survey inevitably presents challenges in terms of register and tone across disciplines, and on this point it seems best to adhere to the terminology appropriate to each domain, in the belief that multidisciplinary approaches—a variety of maps—can take us that much closer to our goal.

Legends and the Scales of Narratological Maps

A fundamental tenet of narratology is that "narrative" and "story" are not necessarily synonymous. Narratologists have made a distinction for years between *narrative discourse*—how listeners/readers encounter events in the order the storyteller chooses to deliver them (sometimes called *sjuzet*)—and *story per se*—the underlying chronology, the timeline of events as they would actually have occurred (*fabula*).[1]

These two—narrative discourse and story—can be very different. They may run "in phase" or "out of phase," depending on the circumstances of the telling, its style, goals, and audience. Think of the novel *Time's Arrow* (1991), by Martin Amis, in which the story of a life is told from its end backward, inverting cause and effect in the process, or *Memento* (2000), a film that gives us the narrative of an amnesiac struggling (using tricks) to work backward to a forgotten trauma, a desperate daily pursuit of the self repeatedly witnessed by the viewer and interleaved with flashback sequences.

The field of narratology is a rich and thriving one but is not the main focus here. Rather, the aim is to examine emerging insights into narrative, storytelling, and story reception at least partly from outside the frame of traditional literary criticism, and to summarize and evaluate some of the analytical approaches inspired by neuroscience in a way that is accessible to non-specialists. Our journey will traverse, at various times, the fields of cognitive poetics, cognitive rhetoric, Darwinian literary criticism, and linguistics, as well as so-called hard neuroscience: a daunting itinerary perhaps, but, one that often contains ideas that are eminently graspable, often demonstrable, and always revealing.

For our purposes, a broad working definition of narrative will be used— one accepted by theorists such as Seymour Chatman and Mieke Bal—as *the representation of an event or series of events*. This allows the greatest latitude for an investigation of biocultural, cognitive, and neuroscientific aspects. As regards fiction, it would encompass historical forms as diverse as myth, epic, legend, saga, romance, allegory, chronicle, satire, drama, short stories, and the novel. Yet the term "narrative" is much larger than story. It also relates to some non-fiction—even many disciplines traditionally seen as scientific.

Story is ubiquitous. In the words of Richard Kearney (2002), "Telling stories is as basic to human beings as eating. More so, in fact, for while food makes us live, stories are what make our lives worth living. They are what make our condition human" (3). That is to say, the act of storytelling is a cognitive channel for meaning-making, allowing us to make sense, not just of our environment, but of our lived experience as it unfolds over time. Stories are, in other words, profoundly useful instruments for understanding the world we find ourselves in, "holding a mirror up to nature," and discovering ourselves in the process: an interpretative screen we both look through, and see ourselves reflected in. Kearney elaborates, "Every life is in search of a narrative. We all seek, willy-nilly, to introduce some kind of concord into the everyday discord and dispersal we find about us. We may, therefore, agree with the poet who described narrative as a stay against confusion" (4).[2]

Psychologist Jerome Bruner, in a much-cited 1991 paper, also sees deep-seated processes behind our narrative drive: "We organize our experience and our memory of human happenings mainly in the form of narratives—stories, excuses, myths, reasons for doing and not doing." Like Kearney, he points out that the relationship between language and thought is one of feedback and reciprocity. "Cultural products, like language and other symbolic systems, mediate thought and place their stamp on our representation of reality." Describing the dual role of narrative in thought and discourse, Bruner uses a bold medical metaphor, "as with all prosthetic devices, each enables and gives form to the other, just as the structure of language and the structure of thought eventually become inextricable" (1991: 4 and 15).

The implication is clear—narrative discourse and narrative thought are at once prisms through which we view events, and practical, organizational aids for negotiating an uncertain world. The use of "prosthetic device" as a metaphor raises a profound possibility: like such devices, can stories become assimilated into—or are they already part of—our core identity? Do we narrativize not just the world, but ourselves, and ourselves within the world? Neuroscientist Antonio Damasio certainly believes so: "*Subjectivity is a relentlessly constructed narrative.* The narrative arises from the circumstances of organisms with certain brain specifications as they interact with the world around, the world of their past memories, and the world of their interior" (2018: 159, his italics).

Evidence for our inherent "narrativity" comes from many directions, but the notion is not without its detractors. One of the first points this book addresses is the extent to which narrative is a human universal, beginning with campfire tales, and traveling toward support from brain science and allied fields.

Narrative Universality

Of course, there is some distance between oral storytelling and literacy, the common thread being our use of language. Writing emerges from talking. But then, as E. L. Doctorow's character Andrew (*Andrew's Brain*, Chapter 10) remarks to his therapist when pushed to keep a journal, "Writing is like talking to yourself, which I have been doing with you all along anyway, Doc" (2014: 49).

A prominent voice in cognitive literary studies, Patrick Colm Hogan (2003a) begins his take on the role of narrative by accepting that storytelling "is an activity engaged in by *all people* at *all times*" (3, our italics). Consequently, he says, empirical approaches to the study of mind are actually doomed to fail if they do not engage with this most compulsive of human behaviors; "cognitive science can hardly claim to explain the human mind if it fails to deal with such a ubiquitous and significant aspect of human mental activity as literature" (4).

That narrative behavior is found in all human groups may initially appear to be a trivial insight. Hogan suggests the reason it seems unremarkable lies in a paradox related to its very ubiquity. In some ways, this is akin to the problem pointed out by Noam Chomsky in relation to language universals. It is hard for us to see the need to explain everyday phenomena to which we have been familiarized from our earliest years. Similarly, Hogan suggests, as speaking and thinking beings, we tend to overlook the need to explain our abilities to speak and think. So, too, with our storytelling compulsion: it is "hiding in plain sight." We need to remind ourselves that when it comes to literature, "There is no logical necessity in the existence of [linguistic] art, and in our own society, very few people actually produce it. Why then should we expect it to appear in every society?" (22).

Hogan developed a historical research program looking at literary forms across cultures, his particular interest being the centrality of stories as they, first, represent and then give rise to emotion in readers. As he points out, stories that recur frequently (what he calls "paradigm" stories) are structured by emotions; they are affecting. Indeed, you might argue they come to be regarded as great literature partly to the extent that they are able to "move" successive generations of readers. Hogan sees an evolutionary basis for this, pointing to the universal facial expressions associated with typical emotions, in a description Darwin would have recognized. The notion that emotional experience or expression is itself universal we will come to from a neuroscientific perspective in Chapter 7.

Furthermore, he sees the source of many emotions in narrative self-evaluation, relative to goals we all understand—getting the girl/boy, achieving social status, competing, and winning. This is the story as *quest*. "Keith Oatley and Philip Johnson-Laird have argued persuasively that emotion is the product of an agent's evaluations of his/her success or failure in achieving particular goals within what is, in effect, a narrative structure" (76).

Hogan, Oatley, and many others propose that our ability to emotionally relate to and empathize with fictional characters is based, above all, on our own autobiographical memories. Yet, as we shall see, these memories themselves may well have been structured and laid down according to narrative principles. It is yet another example of dynamic reciprocity at work.

Universals and Particulars

Expanding on the points above, Hogan not only argues for the production and reception of story as a *human universal*, but takes the much stronger position that certain story *forms* are too. The kinds of universal story schemata he points to are also goal-driven, and include, for instance, heroic epics, romantic comedies, power struggles, and stories of sacrifice. Specifically, "romantic union and social or political power ... are the two predominant prototypes for the eliciting conditions of happiness. Thus, they are the prototypical outcomes from which our prototype narratives—including literary narratives—are generated" (Hogan 2003a: 94).

One implication of Hogan's belief in universal literary forms is an acceptance of the shared nature of emotional experience, and even expression, across peoples. Hogan takes one further step to suggest—and again, it's a big claim—that the commonality of emotional narrative strongly suggests the narrative roots of human emotions.

Inclusive Cognition

Distinguishing cognition from affect was a large part of Descartes' philosophical agenda: reason and emotion were seen in opposition. His stance echoed much older views, like those of Plato, that emotions often conflict with reason, perhaps one reason why he banished poets from his idealized "Republic."[3] An antagonism between cognition and emotion thus became enshrined in Western philosophy. Over the years, fiction has thrown up its fair share of "Dr. Spocks" or "Gradgrinds": characters who embody reasoning devoid of feeling or emotion. Yet the science of the late twentieth and twenty-first centuries clearly points to the distributed nature of cognition in our brains, and to functional overlaps with affective states and emotions (see Chapter 7). In some ways, this should not be a surprise—emotions are an evolutionarily ancient and adaptive phenomenon. There can be no "cordon sanitaire" between cognition and affect.

A well-considered and nuanced position on the division between affect and intellect, or feeling and cognition, was articulated several decades ago by Antonio Damasio in *Descartes Error* (1994) where he brought together observations from behavior, brain anatomy, and neurology to show that attempts to make "rational" decisions devoid of emotion are doomed to fail, are indeed pathological. Reasoning, he argues, is deeply involved with the mental operations of emotion. In his subsequent works (2003, 2018), Damasio accords feeling and emotion a central role in the most basic regulatory process common to all living systems: homeostasis. "Feelings and homeostasis relate to each other closely and consistently. Feelings are the subjective experience of the state of life—that is of homeostasis—in all creatures endowed with a mind and a conscious point of view. We can think of feelings as mental deputies of homeostasis" (2018: 25). Little wonder, then, that they cannot be filtered out, abstracted from, or contrasted with "knowing, perceiving or conceiving" and the supposedly "higher" functions of cognition. Emotion and its conscious correlative, feeling, are part of an ancient life-support system. "Most every image in the main procession we call mind, from the moment the item enters a mental spotlight of attention until it leaves, has a feeling by its side" (2018: 100).

Such an integrative view of emotion and cognition has received additional support from Joseph LeDoux (1996) and, more recently, Lisa Feldman Barrett (2017). As Barrett puts it, "The distinction between emotion and cognition hinges on their alleged separation in the brain ... but ... thinking and feeling are not distinct in the brain ... your brain is always a whirlwind of parallel predictions that compete with one another to determine your actions and experience" (222). Cognitive operations *and* emotions share the vital function of preparing us to act.

In a recent summary, neuroscientists LeDoux and Damasio (2013) elaborate upon that secondary distinction alluded to earlier—between *emotions* and *feelings*. According to them, emotions involve brain changes in arousal along with bodily shifts in endocrine, autonomic, and somatic motor responses. But these are often below conscious awareness. However, "the term feeling [refers] to the conscious experience of these somatic and neural changes. In a certain sense feelings are *accounts our brain creates* to represent the physiological phenomena generated by the emotional state" (1079, our italics).[4] This will also be discussed more fully in Chapter 7.

Reading Mental and Emotional States

As we are highly social creatures, a biocultural view of our needs suggests that an ability to interpret and anticipate the thoughts, intentions, and emotions of others would be a massive advantage. Indeed, most of us do exactly that for much of the day: an instinctual act known as "mindreading." Indeed there is considerable evidence of an adaptive value, linked to status and cooperation, in projecting ourselves into a fictionalized engagement with the minds of others in this way.

Our capacity to mindread can be seen in children from the early years, and even, to an extent, in other species, such as dogs. Habitual and barely conscious mindreading (or—as it's also known: mentalizing or, confusingly, "theory of mind") can seem so banal as to be barely worthy of comment.[5] Yet, as with our hunger for stories, the existence of an almost universal mentalizing faculty is actually quite remarkable, given the fact that we are each trapped in our own perceptual "box," and possess no direct sensory means to apprehend the interior life of others.

Strictly, we can only directly experience *our* perceptions: private, inner "qualia," as they are known. In Hogan's words,

> Since I do not have direct access to the "qualia" of another person ... how can I judge that the two of us are feeling the same thing? ... Stanley Schachter's work is particularly relevant here. Schachter has argued that an emotion involves some state of physiological arousal plus some cognitive interpretation of that state ... It is the cognition which determines whether the state of physiological arousal will be labeled "anger," "joy" or whatever ... knowledge of the self is produced by the same strategies as knowledge of other social objects, and is prone to essentially the same sorts of bias and error. (2003a: 240)

It may be for precisely this reason that story holds up not only a "mirror to nature" but a mirror to ourselves, providing archetypes that we internalize and use to label our own emotions at an early age, later integrating them into an account of self (Eakin 1999, Bruner 2004).

Another suggestion is that stories provide us with a kind of phantom-realm where we can engage in low-risk "what if" scenarios—offering the opportunity for "shadow-play" with a virtualized self. The advantages in terms of future planning and hypothetical run-throughs of possible happenings seem obvious.

Narrative, History, and Natural History

So central is narrative to human interaction and engagement with the world that Hayden White, in *The Content of the Form* (1987), pointed out that its etymology can be traced all the way back to the ancient Sanskrit "gna," a term that means "know." Hence, it is unsurprising that a huge amount of information has been amassed about stories across ethnic and geographical divides throughout human history. From a biological perspective, stories appear as a human universal and therefore a natural phenomenon, whereas the anthropological perspective addresses them as cultural. Again, there is no inherent conflict in this.

Storytelling/listening is one of many areas where biology and culture interact, and increasingly research is being done across what used to be known as the nature/nurture divide. For these reasons, a focus on story is more than an obvious way for science to relate to the arts. Understanding the intrinsic "narrativity" of our minds can open up a view of how science is initiated in the act of hypothesizing, structuring a narrative, and then communicating it to a wider audience. In the coming chapters, we shall enquire: what mental faculties are engaged by stories, what narrative techniques mediate this, using which brain mechanisms. We will also ask whether the same faculties and narrative predispositions come into play during scientific inquiry.

Value Propositions

A long-standing proponent of a strong "narrativist" perspective, Roger Schank (1995) put forward several compelling reasons for the importance of story, "Knowledge is … experiences and stories, and intelligence is the apt use of experience and the creation and telling of stories. Memory is memory for stories, and the major processes of memory are the creation, storage, and retrieval of stories." Or, even more radically, "Storytelling and understanding are functionally the same thing" (16).

David Herman (2013) reinforces this core utility of stories. In his words, "narrative gives scaffolding for making sense of experience," using the example of someone who has been subjected to an act of violence (88). Such traumatic circumstances create a pressing need to "make sense" of what has happened and to allow for what has become known as "closure." A traumatic accident is rendered comprehensible largely through its many narrative re-workings: in, for instance, police reports, lawyers' accounts, as well as those of therapists or counselors, friends, family, etc.

His example underscores the point that narrative is not something rarefied or abstracted from ordinary living. It is a framework to help us break down and assimilate the randomness of events. Consider the worn cliché of someone's "battle" against cancer. This is metaphorical narrative used as both shield and solace.

Before leaving Herman's account, we should outline a vital distinction made by him that also runs through this book. Over the years, one concept that Herman has written extensively about has been that of the "storyworld." This term has been used by narrative psychologists for some time and in Herman's sense it is a world evoked by a narrative, and completed by the listener/reader, irrespective of whether it takes the form of a book, film, graphic novel, or everyday conversation. He describes the storyworld as an emotionally engaging realm in which a reader feels invested, and which is made up of an associative field of memory, imagination, and emotion. Furthermore, he argues, convincingly, that stories act as scaffolds, or armatures for the ascription of intentions to characters. They are intentional systems, albeit provisional. Like maps, stories are structures which require the participation of the reader (or "percipient" in cartographic terms). They omit far more than they contain, and require our powers of implication and association to fill in the gaps.

Where his analysis becomes most useful is when he juxtaposes two related ideas: The first is fictional narrative as a semiotic game with its own rules and codes. As reader, we enter into and animate this make-believe world on the principle of minimal departure (Ryan 1980) from the nature and causality of our own daily actuality. Only then can we reach the enticing stuff—what's happening, to whom, where, and why.

The second is the application of those same narrative techniques projected outward onto that very same actuality, to make sense of the world (rendering the accidental, coherent). In the first, we are looking at what he calls "worlding the story," in the second "storying the world."

Worlding the story relates to the creative act. It mainly refers to the process of bringing the story to life—a cooperative venture between writer and reader. For the former, this is about constructing an internally consistent intentional framework with enough "actualized" elements to permit our vicarious participation, grounding the illusion.

Storying the world, on the other hand, is something we all do, using narrative structures to make sense of experience, configuring our daily circumstances, often random or meaningless, into coherent scenarios using the devices and strategies of story. This may be "implicit and unconscious," and vary in extent between individuals. At the very least, fictional narrative offers the chance to playfully practice mindreading skills and impute causal relations—both processes offering "real-world" advantages.

The Clever Woman of 1001 Stories

In *1001 Nights*, Scheherazade found herself newly wedded to a king—Shahryar—so wounded by infidelity that he vowed to take a new wife each night and assassinate her by daybreak. Scheherazade's nocturnal answer to this dilemma was to weave tales so compelling that the king was desperate to hear the ending. But by the time dawn breaks each day, the story is never complete. The ultimate cliff-hanger.

One can see why Mark Turner in his seminal work of cognitive literary criticism, *The Literary Mind* (1996), reaches for the Scheherazade parable to begin his study of metaphor and conceptual blending. It is, after all, the archetypal story about the value of story. In the context of this book, Scheherazade is important for another reason. She not only is the storyteller, but turns up as one of the most significant characters in the frame narrative of *One Thousand and One Nights*. Indeed, her tales contain other frame narratives (embedded stories told by one character to another). The collection is thus itself a prime example of *recursion* in literature—a feature we will return to. Within the bounds of the dominant frame narrative, the king's need for plot resolution, for finding out what happens in the tale, spares Scheherazade to live another day. Story is her lifeline.

There are many who argue that it is ours too—that without internal narrative a person may not have a coherent "self" at all, where others, like Damasio, see even momentary consciousness as an ongoing story, the ultimate reality television show we tell ourselves. With this difference, the "show" generates the self. Whatever way you cut it, there appears to be something so fundamental about basic story structure that in any study of human minds, the role played by narrative, together with our universal hunger for it, demands examination. For we are perpetual Scheherazades, ceaselessly deferring the moment when the stories stop (see Chapters 9 and 10).

The (Word) Play's the Thing

There are many aspects of ordinary language that shed light on what stories do and how they do it. One tool that's essential to literary imagination is metaphor. Keith Oatley is one who has considered it throughout his career; in a recent work (2011) making the eminently practical point—one we will encounter in varied ways ahead—that part of the writer's job is algorithmic, using language, in particular, metaphor, to cue us (or *instruct* us) to imagine a particular scene, or feel a particular way at a precise juncture in a text.

Taking his lead from Shakespeare, Oatley begins with the notion of the fictional realm as a dream, or virtual play space, which he elaborates as "metaphor in the large." The ancient term for such extended imaginative pursuit was *mimesis*, the more modern equivalent being *simulation*. In his words, "Narrative stories are simulations that run not on computers, but on minds" (17). Yet he is also careful to emphasize that entering such narrative simulations is very much an active process of engagement and reconstruction, rather than passive consumption. This is where films, computer simulations, graphic novels, hypertext novels, and even dramas differ from conventional prose. Though adhering to many of the rules of narrative, through their graphic plentitude, such platforms remove much of the imaginative work of internally "mapping" the fictional domain. We still have to do much inferential work, and often much "mindreading," but we can see the lay of the land.

Of course, some of these other forms also allow interactivity, in a way that is only slowly and hesitantly entering the more traditional narrative forms. The playwright Alan Ayckbourn amongst others has experimented with interactive drama, hypertext novels allow readers choices which subsequently influence plot outcomes, and recently, the "Black Mirror" episode of *Bandersnatch* on Netflix incorporated viewer options at a number of pivotal points in the plot to activate a decision-tree of narrative possibilities. Arguably, entering a storyworld through the non-pictorial portals of prose or poetry involves a greater level of attention, and effort, if not empathy.

Oatley (like literary Darwinist Brian Boyd) ultimately sees the source of fiction in play, especially symbolic play, or more simply, pretending. Why so? Well, symbolic play at its simplest involves the ability to see one thing as another: to take a stick and wield it as a sword, to throw an apple as a ball. Perhaps seeing some superficial similarity between two objects, the child, in a connective leap of imagination, turns "A" into "B," and now quite unselfconsciously relates to the transformed object in its new capacity. We have entered the world of pretense, or what Fauconnier and Turner call "counterfactuals," whereby people "imitate, lie, fantasize, deceive, delude, consider alternatives, simulate, make models and propose hypotheses" (McConachie 2007: 568). This is a world familiar to artists, and especially storytellers, but also, with respect to the last four of these activities, to scientists.

The child who engages in such play has accessed, without knowing it, the almost magical power of metaphor, and now can enter what might be called the *multivalency of the mundane*. The basic discovery in a remote past that anything can be both itself and something quite distinct (or even a number of things) must have been a startling and liberating revelation, further heightened by the realization that such overlap can be utilized in storytelling and anecdote. Turner in particular gives a detailed account, under the heading of "conceptual blending" of some of the mental operations involved (see Turner 1996, Fauconnier and Turner 2002).

Researchers of prehistory—among them Steven Mithen—view the ability to map one categorical domain (part of a tree) onto another (slashing weapon) as one of the intrinsic aspects of being human. Mithen (1996) characterizes basic artifacts such as shells used as beads, from as early as 82,000 years ago, as displaying our "metaphorical" thinking. These shells had holes bored in them, so we know their owner was able to initially envisage, and then realize their new decorative use. In other words, an *imaginative superposition* occurred. The shell was seen in a totally novel context, then manipulated to make it better fit that context. Others find this argument less convincing, citing decorative behavior in chimps and bower-birds as counterexamples.

Whatever the truths of this, it cannot be denied that metaphors, by their compound nature, are also inherently economical—they save descriptive time and effort. At a mundane level, we witness daily examples of this metaphorical pithiness, for example, a woman describing her (distracted) boyfriend as having "eye fog." They can display considerable complexity. Compound metaphors can show internal incoherence ("mixed metaphors") and become recursively unstable (road sign seen in the UK: "*warning: soft hard shoulder*"). Metaphor is pervasive, indeed, is another of those features so woven into the fabric of our lived experience it can become almost invisible.

A remarkable resource in the consideration of metaphor is the University of Glasgow Metaphor Map (see https://mappingmetaphor.arts.gla.ac.uk/). This vast undertaking charts "all knowledge in English: every word in every sense in the English language for over a millennium" together with their metaphorical connections. It logs 14,000 such metaphorical connections, sourced from 4 million pieces of data, from as far back as 700 AD. The principal investigator on the study, Dr. Wendy Anderson, after three decades of study, concluded that "metaphor is not simply a literary phenomenon; metaphorical thinking underlies the way we make sense of the world conceptually. It governs how we think and how we talk about our day-to-day lives" (article in *The Guardian*, June 30, 2015).

In this broad use of the term "metaphor," it is undeniably imaginative work, a generative function not just for the high-level processes involved in "creative" pursuits such as narrative construction, but also for the richness of our social interactions, our understanding of the external world, and even our own cognition. It is no accident that the three-color-coded categories into which the thousands of metaphors in the Glasgow map naturally fall are the external world, the mental world, and the social world. The word itself derives from the Greek *metapherein*: to transfer, and necessarily entails the sense of movement. A world devoid of transformative metaphorical "play" would be inert, lacking an entire cognitive dimension.

There has been extensive analysis of metaphor and the process of taking something familiar from its natural state (the source domain) and reapplying it in a new, less familiar context (the target domain). So, for instance, the experience of my hand grasping an object is projected onto the more abstract notion of "grasping" a concept. As Oatley puts it, "In Lakoff and Johnson's sense [thought] consists in applying features that we do understand in the source domain to things we seek to understand in a target domain" (33).

From this claim, seemingly innocuous, but actually huge in its implications, it is not a long way to the ideas of Turner who identifies metaphor as a requisite of both story and thought, clearly starting from a "narrativist" position, in keeping with some of the other thinkers just described: Bruner, Damasio, and Kearney. Turner (1996) puts it concisely: "Narrative imagining—story—is the fundamental instrument of thought" (4).

Oatley concurs that metaphor plays a key role in meaning-making, especially as regards the self, summarizing Turner's description of the two steps involved: "The first is to form a story, a sequence about what someone did and what events occurred. The second is to project this story onto another story, for instance, onto the story of one's own life" (2011: 33). Setting aside any such autobiographical claims for a moment, metaphor is indisputably a key building block of the literary imagination, and one that neuroscience is beginning to probe.

Imagination Dead Imagine

Surprisingly, human imagination has only recently become a "respectable" process for neuroscientific research. In fact, you might date its recent scientific emergence to the year 2007, a time when studies of the newly defined "default mode network" began to show the neurological link between memory and imagination (see Chapter 8). Of course, in the culture at large, imagination has been a perennial subject of fascination and debate, engaging philosophers, psychologists, critics, and anyone who has found himself or herself disappearing into a tale.

Traditionally, as its roots suggest, the word "imagination" had a visual connotation. However, partly thanks to neuroscience we now have convincing evidence that our definition of what constitutes, or may constitute, an "image" must be expanded to include auditory, tactile, somatic, and multisensory representations. Given this more sophisticated multimodal understanding of what the term might mean, now is surely the right time to ask, what sort of mental representations would stories trigger? An off-the-cuff answer would perhaps be an auditory one: the inner voice. Certainly the concept of an "inner voice," and its link to the faculty of imagination, has been obsessively explored by writers over the years, including Virginia Woolf and Samuel Beckett: "A voice comes to one in the dark. Imagine" (the opening from Beckett's novella *Company*, 1979). From a critical perspective, too, the inner voice has an immediate appeal because it offers a bridge between the primary story modes: oral narratives (mostly auditory in origin) and written narratives (which originate as visual, linguistically bound information).

Samuel Beckett's work demonstrates a long-term fascination with the mental processes behind storytelling and the relationship between the heard and the read, whether in his instruction to the actor Billie Whitelaw to deliver his play *Not I*, at the "speed of thought," his exploration of the strange recursive paradoxes offered by language (as in the title of this section—one of his short prose pieces) or even his late expression of interest in neurosurgery, and his extraordinary admission that "I have long believed, that here, in the end, is the writer's best chance, gazing into the synaptic chasm" (Shainberg 1987: 2).

In *Krapp's Last Tape*, Beckett takes the incessant inner stream of "self-talk," fragments it, and, exploiting Krapp's "narrativist" need, externalizes the process, showing him "piecing himself together" by listening to old recordings of himself.

Famously, in regard to the inner voice, Noam Chomsky suggested that by far the most common use of language is not interpersonal communication at all, but talking to oneself. The latest findings in experimental neuroscience give some weight to Chomsky's insistence on the primacy of inner vocalizing. Imaging studies using PET and functional MRI (see Chapters 4–6, and the Appendix) have shown that many of the same linguistic regions of the brain are activated when we listen to words, view words, or simply call them to mind. But knowing that a few regions of the brain are activated in this way does not tell us *what is going on* in them: this requires observations at a neural level to determine what information is being called up or exchanged among parts of the linguistic system.

In an ingenious experiment, a team, including Brian Pasley, Stephanie Martin, and colleagues (2012, 2014), used intracranial recordings to try and detect speech "from the inside." To do this, they placed electrode arrays directly on the cortical surface of patients who were being assessed prior to epilepsy surgery. The patterns of electrical signals recorded from "non-primary" auditory cortex could be correlated with the acoustic features of words presented to the patients (see Chapter 4 for a primer on relevant brain areas). Thus the team was able to develop algorithms for predicting the identity of particular words heard by subjects using *only* the cortical electrical activity that was evoked. Amazingly, after a time, their techniques allowed them to identify patterns of not only overt speech in the brain (words being heard) but also covert speech (words being formed during silent reading).

In effect, they used native electrical signals from a specific language region to reconstruct the "sound" of one's inner voice. Based on this ingenious research, we can say with some confidence what neurons in certain areas of auditory cortex are up to. In part, they are encoding the phonemic (acoustic) properties of words as we think—and so they may be part of the mechanism that ties thought to our subjective sense of heard speech.

This suggests some sinister possibilities, but also, in a medical context, some potentially wonderful applications for patients unable to speak, if we can reliably "read" their thoughts or intentions by assessing brain activity directly. Think of those with "locked in syndrome," where a person is almost completely paralyzed and largely unable to communicate, yet has intact cognitive functions. Such was the case of Jean-Dominique Bauby, who had to dictate his autobiographical account, *The Diving Bell and the Butterfly* (1998) through a strenuous communication system based on eye movements, his only remaining motor capability (see Chapter 6). How much more we might have heard from him if we could have eavesdropped upon the inner discourse of his imprisoned, but vibrant imagination.

However incredible the recording of covert vocalizing may seem in itself, it also leads us into many new areas of inquiry. The novelist Robertson Davies (1990) went so far as to make the case for developing one's inner voice to enhance the reading experience: "Are we not foolish to give up that inward voice in which books can speak to us? … [an important aspect] of reading is listening to the inward reading voice" (16). He even recommended that we should hone the ability to adapt our inward voice during reading so that, for example, different characters in a novel would develop unique accents—imaginary "idiolects."

Annie Dillard (1989) addressed the related phenomenon of the inner ear in the act of creative writing when she said:

> This writing that you do, that so thrills you … is barely audible to anyone else. The reader's ear must adjust down from loud life to the subtle, imaginary sounds of the written word. Any ordinary reader picking up a book can't yet hear a thing; it will take half an hour to pick up the writing's modulations, its ups and downs and louds and softs. (17)

Dillard's way of connecting the inner ear with our effort to hear the "imaginary sounds of the written word" makes it evident that the potential ability of neuroscience to record our inner voice is also the potential to document the reality of *one* form of literary imagination. Whether appealing predominantly to our inner ear or inner eye or any other of the senses (Chapter 6), perhaps it is little wonder that we are all, like Scheherazade's King Shahryar, held, captivated, and beguiled by stories.

Notes

1 A *narrative text* is one in which an agent conveys to an addressee ("tells the reader") a story in a particular medium, such as language, imagery, gestures, or their combination. For more on *fabula, sjuzet* see Mieke Bal (2009), for an introduction to the Russian formalists see Peter Barry (2002).

2 The poet referred to here is Robert Frost.

3 O'Grady (1995) entry on emotion and feeling, *The Oxford Companion to Philosophy*, p. 225. T. Honderich ed. In Plato's famous analogy, reason is like a charioteer struggling to control two horses: emotion and physical appetites.

4 Joseph E. LeDoux and Antonio Damasio (2013) *Emotions and Feelings*, chapter 48 in Kandel et al., 5th edn.

5 Mercier and Sperber (2017, especially chapter 6) have discussed the implications of using the terms "mindreading" versus "theory of mind." We rely mostly on the former term.

2

Orchestrating the Imagination

Elaine Scarry's *Dreaming by the Book* ([1999] 2001) starts with an anecdote that goes to the very heart of our experience of story.

> In the poem "Birthplace," Seamus Heaney describes a young boy named Seamus Heaney staying up all night to read for the first time a novel (Thomas Hardy's *Return of the Native*) and at dawn not knowing whether the newborn sounds of bird and rooster and dog were coming to him from the surface of the field or from the surface of the page. (6–7)

Prompting her to ask, "by what miracle is a writer able to incite us to bring forth mental images that resemble in their quality not our own daydreaming but our own much more freely practiced perceptual acts?" (Scarry 2001: 7).

There is a lot packed into this sentence: the relation of perception to imagery, the complex question of what comprises a mental image, and the mysterious nature of mental states evoked by the authorial choice of words on a page. There are ways to think about these issues focusing primarily on literary devices and usage, utilizing the long-established terminology of symbolism, tropes, pragmatics, and stylistics. What is increasingly apparent is that there are also ways to think about them in regard to neural function.

Intuition and Perception

That we have an "inner voice" which can be cued by reading seems attested to by common parlance and phenomenological experience: we speak of the "sound" or "rhythm" of a writer's prose or poetry. In the course of our normal life, we are habitually party to self-generated inner monologues, extended "conversations," and indeed rehearsed arguments, as well as inner music—"tunes in the head." In a 2020 study of 181 published writers attending the Edinburgh Book Festival, two thirds reported hearing their character's voices (Foxwell et al. 102901). But introspective realities can be deceptive. Science is full of examples where widely held convictions are found to be groundless when careful evidence was sought.[1] Brain science has now independently verified the occurrence of such auditory traces—"soundless voices" as E. L. Doctorow dubs them, when we read, or simply ruminate. Even more remarkable, it has become possible to decode some of these unspoken vocalizations ("thoughts") from the outside with some degree of accuracy (see Chapter 1).

The act of reading is not of course restricted to inner vocalizing. For example, many narratives, especially literary ones, are rich in sensory description, and we speak of them bringing scenes "to life," suggesting they activate our own perceptions of sight, smell, taste, and touch. Take James Joyce's famous description in *Ulysses* of the sea at Sandycove, Dublin, "the sea ... The snotgreen sea. The scrotumtightening sea" (Joyce 1922: 11). It works simultaneously on several levels, which we might call the melodic and the harmonic. In the first, we glean the meaning horizontally, traversing the sentences sequentially, being forced to pause and recalibrate our understanding as we absorb the playfully carnal "rightness" of the compound adjectives. But there is another level at work too, heavily hinted at in context, harmonic and allusive, referring us to the sea as depicted by the poet Swinburne, and even further back, to the Greek, Xenophon. This connotative axis is vertical, associative, reaching into historical (and rememorative) time, and operates on us like an underlying chord. Reading, then, can be multimodal, polyvalent, and diachronic.

We often share and enjoy stories which successfully use corporeal triggers whether gustatory, olfactory, or kinesthetic, to activate vivid memories—Proust's "madeleine" being the most famous example. Nevertheless, it appears to be the aural and visual modes (what Joyce called the "ineluctable modality of the visible," 1922: 12) that dominate the creation and reception of story, helping to elicit our empathetic investment in entirely invented personalities and their fates. However, "the reality represented in the text is not meant to represent reality; it is a pointer to something that it is not, although its function is to make that something conceivable" (Iser 1993: 13). To put it another way, "storyworlds are possible worlds" (Palmer 2010: 179), or in Peter Turchi's memorable reformulation, "a story or novel is a kind of map because, like a map, it is not a world, but it evokes one" (2004: 166).

In what follows, a small selection of authorial techniques for this imaginative mapping—the conjuring up of "storyworlds"—will be addressed. Of special interest is the stimulation of sensory modalities and images, but also the question of how such textual triggers can lead to emotional engagement, and even the readers "transportation" into the narrative realm (à la Heaney). By looking at storytelling techniques (tricks?) from a neuroscientific perspective, we can now highlight how they might achieve the effects they do.

Down the Rabbit-hole

There is a question every reader asks themselves in the first few sentences of any story, and every writer is desperate to answer: why should I care? In emotionally driven narratives, such as novels, the imperative to entice the reader "down the rabbit-hole," to enter the illusory "storyworld," inhabit the characters, their actions, senses, fears, and aspirations, is paramount. For this purpose, the authorial manipulation of what we "see" and what we "hear" in the fictional realm, and how we "hear" and "see" it, is crucial. It might be said it is this multisensory ability to inveigle us over the border into terra nova that is one mark of major writers from Cervantes to Austen to Virginia Woolf. In order to do this, they often invent a guide and wayfinder, someone who will stay close at hand to lead us into the unknown. First-person narration involves an even more intimate relationship, akin to an act of possession. I "become" my own guide—their eyes: my eyes, their ears, mine.

Novelist and critic David Lodge noted something, on the face of it so obvious it is often lost: "Whatever the novelist ... does *qua* novelist [is done] in and through language" (Lodge 1966: 164). Taking this reminder in the spirit in which it was given, we will look at some of the tools available to novelists as, in the service of story, they control our imaginative proximity to their characters and guides. This is sometimes referred to as "narrative distance," and is related to the narrator's—and by implication our—degree of involvement in the story told. There are some analogies here to what a film director does, zooming from a "long-shot" to a "close-up"—in fictional terms, moving from the cool external reporting of behavior, to, at the other extreme, "inhabiting" a fictionally constructed mind.

Even though as readers we often barely apprehend the authorial mastery involved, the creation of such a route into a characters innermost consciousness is an extraordinary act. In cinematic terms, it is akin to the point-of-view shot, yet in many respects more complete and, of course, achieved entirely through language.

I. A. Richards distinguished two broad functions of language, the *referential* and the *emotive*, the latter carrying with it a greater weight of association and ambiguity. Focusing on emotion will help to foreground how writers communicate and organize the cognitive realities of experience.

In an early exploration of this concept, Leech and Short ([1981] 2007) cite the paragraph below, taken from about mid-way through William Golding's novel *The Inheritors* ([1955] 1988), their analysis drawing on a seminal study of cognitive grammar by Michael Halliday ([1971] 2002). The action, set in prehistory, centers around the interaction between *Homo sapiens* and Neanderthals at a time when the former are wiping out the latter (see Chapter 5). Golding's novel has further relevance here as regards the evolution of language in our hominin forebears—something Darwin himself was fascinated by, and which continues to provoke debate amongst paleoanthropologists and others (Chapter 5).

What makes Golding's novel striking is the fact that the story is related almost entirely from the perspective of the Neanderthals. The term "focalization" is useful in this context, referring to the fictive lens, and more, to the fictive body through which we experience unfolding events. In the words of Porter Abbott (2008), "Just as we pick up various intensities of thought and feeling from the [fictional] voice that we hear, so also do we pick up thought and feeling from the eyes we see through" (74).

The passage quoted below occurs long after the reader has had time to adapt to the strangeness (to us) of the Neanderthal outlook. Even though, by this point, we have been "living with" and sharing the concerns of a Neanderthal protagonist—Lok—and his kin for almost 100 pages, the description makes great demands of us. Because our visual cues are channeled entirely through Lok, and because his perceptions are in some ways enhanced, in others impaired, by what we would see as causal and linguistic deficits, it takes us some time to work out what is going on. When we don these particular VR goggles, we find ourselves with a very foreign and disturbing "Umwelt": a unique sensory world and cognitive outlook.

Golding's passage represents one of those unusual moments in fiction where the optical cues, or prompts, afforded by the writer are quite at odds with—even run directly counter to—our readerly need to make sense of events. For a moment we experience an unnerving disconnect, an aporia. The seamlessness of our imaginative inner projection falters, faced with conceptual lacunae:

> The bushes twitched again. Lok steadied by the tree and gazed. A head and a chest faced him, half-hidden. There were white bone things behind the leaves and hair. The man had white bone things above his eyes and under his mouth so that his face was longer than a face should be. The man turned sideways in the bushes and looked at Lok along his shoulder. A stick rose upright and there was a lump of bone in the middle ... Suddenly Lok understood that the man was holding the stick out to him but neither he nor Lok could reach across the river ... The stick began to grow shorter at both ends. Then it shot out to full length again.
> The dead tree by Lok's ear acquired a voice.
> "Clop!"
> His ears twitched and he turned to the tree. (Golding 1988: 96)

Here, Golding pits the imaginative work of fiction, concerned with establishing, furnishing, and maintaining the "storyworld," against the reader's emotional investment in Lok's point of view: visualization against empathy. For a moment, the "lucid dream" of the fictional world glitches and stalls. It is a dangerous authorial strategy, but a deliberate one.

There is a moment, early on, when the realization hits us: in order to grasp what is going on, to get the "film reel" moving again, we have to acquiesce to an "Umwelt" with a limited notion of human agency. What is fascinating is the way the passage plays with our capacity for "mindreading." Golding here asks us to enter Lok's mind, to inhabit Lok's worldview, yet that worldview itself is the result of a highly deficient mindreading capacity.

In the last few decades, writers of neurofiction have increasingly asked us to enter the minds of protagonists with similarly compromised or pathologized mindreading abilities—Alan Palmer (2010), one critic who has focused on how fictional narrative portrays others' minds, described such characters, particularly those on the autistic spectrum, as exhibiting "mindblindness" (a term introduced by autism researcher Simon Baron-Cohen 1995). An excellent example is Mark Haddon's *The Curious Incident of the Dog in the Night-time* (2004).

In this instance, despite the conceptual paucity of their worldview, Neanderthals appear to have faculties we lack, can wordlessly "share" mental pictures among themselves and have an animizing openness to the natural world. Their fatal inabilities are in grasping motivation, intention, and volition as regards this strange new species: us. By highlighting this, Golding weighs the balance of our own mentalizing—by diminishing our ability to *think* like Lok, he makes it necessary to *feel* more like him—and, in the process, draws attention to human mentalizing as the extraordinarily fluid capacity it is.

In the passage above, Lok has no grasp of the hostile intent of the man across the river. We can see this in the language used to describe his thought. Even the grammar seems oddly constrained, particularly its lack of clauses with a direct object, or clauses with a human subject. So, when Lok uses his limited powers of deduction, it leads him to a tragi-comic misapprehension: surely the man (who has just fired an arrow at him) was "holding the stick out to him," which seems odd because "neither he nor Lok could reach across the river."

Clauses with a subject are the chief means by which we describe events in which people are involved: "Lok saw the man," or "The man raised his bow and arrow." Instead, descriptions from Lok's perspective are full of inanimate subjects or intransitive verbs or both: "The bushes twitched again," "Lok steadied by the tree," "A stick rose upright."

Intransitive verbs relate to no object. A "stick rose upright" is clearly different in imaginative impact to, "The man raised the stick," and even further conceptually from "The man raised the bow and arrow." The last two instances are transitive—someone does something—thus goal-orientated and volitional. We can now see why Golding avoids them in the presentation of Lok's point of view.

The depiction of Neanderthal consciousness also involves an obvious lexical impoverishment. The man had "white bone things above his eyes and under his mouth," and when the stick rose upright there "was a lump of bone in the middle." We feel his struggles to comprehend exactly what he is faced with. Lok's state of bewildered frustration, tinged with a vague sense of his own inadequacy, is rendered all the more acute when his breakthrough moment, "Suddenly Lok understood," turns out to be so absurdly far from the murderous truth.

So Why Do We Care?

Lisa Zunshine (2006), who approaches the issue through cognition, sees such a process of entering into a focalized fictional consciousness as one of the reasons we engage with novels at all:

> By imagining the hidden mental states of fictional characters, by following the readily available representations of such states throughout the narrative, and by comparing our interpretation of what the given character must be feeling at a given moment with what we assume could be the author's own interpretation, we deliver a rich stimulation to the cognitive adaptations constituting our Theory of Mind. (24)

"Theory of Mind" here being mentalizing, or "mindreading."

Zunshine actively pursues the subtleties of mindreading in fiction, and also a related authorial tool: embedded attribution, or recursive Theory of Mind. This comes into play in the depiction of characters who are using their empathetic skills within a social matrix. So, second order "mindreading" is when you (or readers) hold beliefs about what someone else believes, and fourth is when, for example, you believe that I think that you intend that I believe that something will occur. Much of Jane Austen's work derives from the multiple distortions that can arise from just such echo-chambers of mentalization—as do Shakespearean comedies, and most present-day sitcoms. At the very least, though, this can get a little complicated, the Russian dolls becoming ever more unwieldy. When we superimpose too many projections of intentionality, one on top of the other, we can end up with a blur. Daniel Dennett, who first discussed this recursiveness in 1983, thought it could in principle be infinite.

"Caring" Network(s)

The story of how mental architecture is related to social behavior and our facility for understanding minds, narrativized or actual, has taken off in the last two decades. At the mundane level, we usually have little trouble reading the intentions of others, and the brain sciences have produced a proliferation of theories about how we do it.

One strain of this is based on research findings that monkeys possess a type of neuron that becomes active not only when they perform particular movements, but also when they observe other monkeys or people performing the same ones: so-called mirror neurons (Rizzolatti et al. 1996). Some have speculated that humans also possess these, and that we understand what others intend by referring observed movements to an internal repertoire of motor patterns we have at our disposal. However, this hypothesis is not without problems. For an entry into the debate, it is worth looking at Michael Arbib (2012) on how the mirror neuron idea may figure in language evolution, Paul Armstrong (2013) on its relation to sociality and reading, and Gregory Hickok's broad critique from a communication perspective (2014).

With literature, and our strongly social nature in mind, an experiment performed by Robin Dunbar stands out as a particularly clear example of the rewards of multidisciplinarity in beginning to untangle such knotty problems. Contradicting Dennett, Dunbar showed our capacity to perform mentalizing is not unlimited, but instead occurs within a predictable spectrum: mostly between the second and fourth "orders." Using classical psychological testing methods, he observed that "neurotypical" humans do vary, but most of us lose track, "lose the plot," somewhere between intentionality levels 4 and 5 (Kinderman et al. 1998).

A sample of Kinderman and colleagues' early behavioral findings is shown in Figure 2.1. Note that these experiments had a control: the researchers simultaneously tested the ability to remember factual features of a story *unrelated* to intentionality, and established that in these general memory tasks there was no deterioration beyond four items. That is why the memory line is essentially flat, while the theory-of-mind line jumps up—indicating a specific inability to track beyond level 4. What is their conclusion? The limitations on following higher levels of fictional events are confined to mindreading, not a universal limitation on reading memory *per se*.

In 2010, this group of scientists expanded their behavioral work to include simultaneous structural scans of participant's brains (Powell et al. 2010). What they found was a significant relationship between the intentionality skills of individuals and the volume of brain tissue in the prefrontal cortex (PFC) (the next two chapters will provide some relevant details on brain structure and function). Briefly, damage to this part of the brain, lying just behind the forehead and above the eyes, is linked to deficits in social-emotional behavior. Now, for the first time there was evidence that specific areas of the PFC were larger in people with greater abilities at mentalizing.

The size of brain regions can show some adaptability—plasticity—during your lifetime; the thickness of cortical layers may increase or decrease depending upon experience or pathology. Similarly, some cognitive functions are associated with an allocation of "brain space" that may

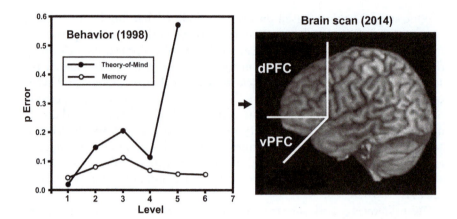

FIGURE 2.1 *The trajectory of experiments on human mentalizing abilities. Data from experimental psychology studies of mentalizing abilities demonstrated during reading.* **Left:** *Rate of errors when questioned about stories that consider varying levels of intentionality (who knows what), or simply factual memory (different levels of non-intentional factual recall).* **Right:** *Lateral view of brain showing two relevant regions of prefrontal cortex (dorsal, dPFC, and ventral, vPFC) that show specific size correlations with mentalizing abilities and also different functional correlations.* **Left:** *From Kinderman et al. (1998) Theory-of-mind deficits and causal attribution.* British Journal of Psychology, 89: 191–204. *Figure 1.* **Right:** *From Powell et al. (2014) Different association between intentionality competence and prefrontal volume in left- and right-handers. Cortex 54: 63–76. Part of Figure 1. Both are used with permission.*

actually increase or decrease in size as behavior changes. The landscape of the brain is subject to change. In this case, connections between the PFC and mindreading were fairly specific. When one sub-region, the so-called ventro-medial PFC, was larger in volume, it predicted that people had greater mentalizing abilities, and even, larger personal social networks (Lewis et al. 2011, Powell et al. 2012).

It is crucial to note that these findings represent correlation, not causality, leaving behind the question: does growth in the size of the PFC lead to greater ability to "read" minds, or does the effort toward better "mindreading" cause growth in the volume of the PFC? Later you will see that in some cases, at least, behavioral and experiential shifts may increase such "brain spaces" (Chapters 7 and 9).

From its earliest beginnings, theorizing in this sector has been guided by an organizing principle: the "social brain hypothesis" (Byrne and Whiten 1988, Dunbar 1998, 2003), which proposes that the large brains of humans came about because of "the cognitive costs of maintaining and servicing large social groups." Undeniably, cooperation within just such groups was key to our success as a species. Such a principle not only leads to the consideration of brain structures, but demands that we factor in "cognitive cost," in time, effort, and attention. Mindreading may come at a price. Dunbar and colleagues have moved in this direction in the latest phase of research, adding functional brain scans into the methodological mix.

With these in play they could observe several indices of "effort" as subjects answered questions about intentionality, or recalled facts in stories, at various levels (Lewis et al. 2017). As predicted, questions about mentalizing required more time to answer than factual detail. At the neurophysiological level, they saw greater neural activation during intentionality questions within PFC and other regions. So the association of a few specific brain regions related to "mindreading" is being ever more rigorously tested; the results provide information that is not only *normative* (what is typical?) but also *particular* (how do we vary?).

Advancing a causal understanding of our social mind is not beyond the reach of empirical analysis, but such a project will take time if it is to play out with proper checks (replications) harnessing all the neuroscientific tools that are becoming available—together with the willingness of humanists, social scientists, and brain scientists to pool their skills.

In and through Language

Before focusing on the basic brain science of reading and writing in later chapters, it's useful, first, to step back and consider the whole issue of the narratability of mind—how writers have developed linguistic devices, tapping into our mindreading skills, for portraying the mental landscape of a character. The most concise way to do this is to very briefly point toward techniques for speech and thought presentation. This is a huge, hotly contested and well-documented topic, addressed in depth by fields such as literary stylistics, cognitive linguistics, reader response theory, and cognitive narratology.

Our intention here is merely to offer the non-specialist an introductory sketch of some of the authorial possibilities available as they relate to fictive mentalizing, and its associated brain states.[2] It is also meant as a reminder that, especially in regard to mindreading and focalizing, "the fictionalizing act is a guided act. It aims at something ... that differs from [our] fantasies, projections, daydreams and other reveries" (Iser 1993: 4).

To begin, we need to consider the conventions involved. In describing conversation, the form most often used is *direct speech* [DS] with an introductory "reporting clause"—"he said" or "she said." So, in the first chapter of *The Inheritors*, we have this account of Lok's tenuous hopes, "We shall find food," he said with all of his wide mouth, "and we shall make love" (Golding 1988: 16). Direct speech like this appears to give a recorded account of the words used by the character, with no authorial interference. We read it almost as if we are eavesdropping on a real conversation. More mediated by the author (and by implication a little less accurate) is *indirect speech* [IS]—"Lok said they would find food and then make love." At once this has a "cooler" emotional hue, and we feel more removed from the character.

But these are far from the only choices. In fact, there is a rich palette of stylistic options available to writers in order to manipulate our emotional and cognitive responses. Such techniques offer the opportunity for subtle shadings of tone, and almost subliminal control of the sense of distance—proximal or distal—that a reader feels in relation to a character. They can also create a convincing illusion of character autonomy. Narratologists have written extensively on these techniques, signaling their central role in facilitating our entry into the imaginative domain (e.g., Cohn 1978, Chatman 1980, Herman 2003, Bal 2009).

David Lodge, as an experienced novelist, is a helpful guide here. Commenting on styles of speech presentation, he notes simply that direct speech is the act of quoting, whereas indirect is the act of reporting. He makes the straightforward point that these two extremes align with the long-accepted distinction between an author "showing" (in this case by quoting) and "telling" (by reporting and interpreting). "I promise to be true," he said, "He promised to be true," and "He swore his fidelity" all have a quite distinctive emotional hues and effects.

The first (direct speech) has often been the default mode in the novel precisely because it lets us "overhear" a character speaking, rather than a narrator's more remote interpretation of the same: it gives the semblance of being, but of course is not, unmediated.

Finally, Lodge (2002) turns to a technique that has been a particular focus of scholarly study and debate, called free indirect speech [FIS] (or, more inclusively, free indirect discourse). This is intermediate between the DS and IS, and emerged during the evolution of the modern novel—dating back at least as far as Jane Austen—to provide the illusion of entry into the interior world of a character: "Novelists discovered free indirect style, which allows the narrative discourse to move freely back and forth between the author's voice and the character's voice without preserving a clear boundary between them" (45). Often, "stream of consciousness" writing, a version of FIS, uses an idiomatic vocabulary and rhythm suitable for the character, reflecting the unpolished language of thought.

FIS generally does not use a reporting clause. It is a hybrid form, allowing a flexible and sometimes ironic overlapping of internal and external viewpoints. Here is an example from Ian McEwan's novel *Saturday* (of which more later) about how Henry Perowne, a neurosurgeon, finds himself losing patients. "Perowne himself is not concerned. Let the defectors go along the corridor or across town. Others will take their place. The sea of neural misery is wide and deep" (19).

So, free indirect discourse is a representation of a character's words or verbalized thoughts that is "indirect" to the extent that pronouns and tenses are aligned with the tense of the character's situation, but "free" to the extent that the thoughts or words quoted appear in the text without quotation marks. The following would be another instance:

"Lok scowled up at the cloud. Yes, it was going to rain."

Here, we move subtly, almost imperceptibly, from a perceptual map to an inner one with the insertion of that word, "Yes."

Much of the potency of free indirect discourse derives from the fact that voice can be rendered in such a way that the speech and thought of writer and character are indistinguishable ("Let the defectors go across the corridor").

The variations and implications of speech depiction are the subject of detailed analysis in literary stylistics.[3] The overall point though is that, even in this apparently simple narrative component, there is a huge range of options available, and each authorial choice has profound consequences for how we perceive the speaker. While supposedly more "faithful" modes like direct speech might seem less interesting when considering character interiority, they do permit us to "snoop" on character conversation as if in everyday life, and to evaluate it for its tonality, emotional undercurrents, and subliminal conflicts: it can also engage our mindreading inclinations.

Command of the range of speech reporting allows writers to, either give the illusion of vacating the stage to their creations, or, instead, stand in the wings with a prompt-sheet and a stage-whisper. Combined with command of the range of *thought* presentation, it offers the potential of transferring us from inside one mental territory (the author's) into another entirely (the character's).

The Inheritors demands undertaking a huge imaginative journey in order to enter the remote regions of the Neanderthal mind. Here, distances are vast: between our inner map and that of the protagonist, and between our outer map and his. The very limits of empathetic "travel" are challenged, prioritizing the entire issue of interiority and exteriority, a fact the novel itself recognizes: "Now, more clearly than ever, there were two Loks, outside and inside. The inner Lok could look forever. But the outer that breathed and heard and smelt and was awake always, was insistent and was tightening on him like another skin" (131).

Unlike speech, where direct reporting is the norm, because we can directly overhear someone's words in everyday life, for the reporting of thought, it turns out that the most common convention (the norm) is what stylists would call *indirect thought*. In a way, this is a tacit admission that we do not have direct access to other people's ruminations.

Indirect thought allows a general flavor of what is happening within the character without committing to precise detail. When a writer chooses to deviate from this norm, by using a "freer" form, we feel that we are crossing a threshold into the consciousness of the character: in the most radical case (*free direct thought*) to the extent of accessing not only their thoughts, but their reactions to their thoughts. Here is McEwan's neurosurgeon Perowne contemplating terrorism, "Is he so frightened that he can't face the fact? The assertions and the questions don't spell themselves out. He experiences them more as a mental shrug followed by an interrogative pulse. This is the pre-verbal language that linguists call mentalese" (80).

To sum up: representations of speech and thought can be exploited by an author in an act of imaginative misdirection. They can "slip from narrative statement to interior portrayal without the reader noticing what has occurred" (Leech and Short: 314). It is even possible to create a "superposition" of authorial and character thought—an ambivalent interiority which can be used to orchestrate the emotions and the imaginative positioning of the reader. The point is there is a sliding scale of narrative possibility here, so much more nuanced than the venerable contrast between third person ("He stood and took his coat.") and first person ("I stood and took my coat.") might suggest.

Before we leave the intimately related topics of mentalizing and focalization (point of view) in fiction, it is worth pointing out that although our empathetic center of gravity as readers is often spoken of as perspectival, it need not reside in a "mind" in the everyday sense at all. We can quite comfortably get *Under the Skin* of an alien, as in Michel Faber's novel of the same name, a ravening monster, like John Gardner's *Grendel*, a zombie, or even a tree. Indeed, "While focalization presents a scene from a particular point in space and time, there is no reason why the mind that occupies this point should be lodged in a body" (Ryan 2010: 486).

Shifting the frame of reference, some raise the possibility that we can identify and empathize with a group mind, a collective mind, since "novels contain a good deal of what is called 'intermental', or joint, group, shared, or collective thinking, as well as 'intramental' or individual thought" (see Alan Palmer 2010: 184, also 2005, on the idea of a "Middlemarch mind" in George Eliot's novel). Critics, together with brain scientists, are looking at our minds in ways that are less individual and abstract, and more and more social and embodied—an issue we'll return to.

By appearing to allow simultaneous access to more than one consciousness, techniques like free indirect discourse have paved the way for a more inclusive definition of mind. Blakey Vermeule (2010)—also asking *Why do we care about literary characters?*—expresses it this way: "Free indirect discourse allows the novel to show consciousness without a first or third-person narrator ... [hence] it sits astride narration and quotation" (75).

An example might help: "David sat and thought about how he'd blown his chances with Sarah. What an idiot he'd been. So clumsy." In these last two sentences, a third-person narrative has co-opted the persona of a first-person voice. This "slippage" is what free indirect discourse allows.

In other words, free indirect thought allows us to hear a character's thoughts in their own idiom, at the same time maintaining a sense of narration (ergo, a narrating presence). Vermeule asks, why *specifically* do authors do this. "They do it to stimulate our mind-reading capacities. To get the two voices running together and apart, a reader has to hold several strands in her mind at once—the sense that several points of view are embodied in each sentence" (77).

In this very brief overview of a few aspects of stylistic choice, we begin to glimpse the palette of options open to the storyteller attempting to reflect the richness of our subjective experience. As the writer wields the palette-knife, subtle (and radical) shifts in focalization and emotional tone and hue can be made to reflect shifts in mental terrain.

Of course, language is involved not only in the portrayal of thought, but in thought itself. The details of how the two are related are still contested. What is now uncontroversial is that stories are indeed "heard" as they are read. Reading, we have the sense of listening to an inner voice. What is wonderful is that neuroscience has been able to confirm this (Chapter 1).

Representation as Re-presentation

In a witty and perceptive "meditation," Peter Mendelsund (2014) delves into what we actually perceive when we read, confirming that our fictional impressions are multimodal: we have an inner voice *and* an inner eye at work while reading, as well as other senses—taste, touch, smell. Moreover, our visual sense of a character almost certainly morphs over the course of a novel and becomes subtly or not-so subtly adulterated, enhanced, or revised by modifying adjectives, the twists of the plot and how we perceive their actions, as well as metaphorical and personal associations and memories.

Mendelsund agrees with I. A. Richards that novels stimulate our imagination by tapping into both perception *and* memory: "Our brains will treat a book as if it were any other of the world's many unfiltered, encrypted signals. ... this is why reading 'works': reading mirrors the procedure by which we acquaint ourselves with the world" (403).

In recent years, critics and brain scientists have been writing extensively about the imagery associated with reading. Some of the most provocative work from the critical perspective has been produced by Elaine Scarry. A. S. Byatt championed Scarry's neuroscience-heavy *Dreaming by the Book* [1999] 2001, claiming, no less, that it proposed nothing less than "a grammar or algebra of the instructions by which a writer causes a mental image to be constructed in the mind of a reader" (Byatt 2006: 251).

Without doubt, *Dreaming by the Book* attempts to lay down the mental foundations for what was historically known as mimesis in literature. This is a massive undertaking. Perhaps unsurprisingly then, the reception to her book was mixed: "We hardly need Scarry to point out to us the fact that authors use their imagination in the process of writing, and spur ours as we read" (Kirkus reviews, August 1, 1999). Such reactions can be seen as yet another incarnation of the standard rejoinder to many of the contributions from both neuroscience and cognitive science: "Don't we know this already?" Again, the short answer is—no, or certainly we didn't when Scarry was articulating her ideas.

As Alan Richardson has pointed out, questions about seeing with the mind's eye, and whether such imagery is related to actual visual perception, have "provoked debate among theorists of literature and aesthetics from the eighteenth century to the present. The same questions, however, have inspired a good deal of (sometimes heated) argument in the cognitive sciences as well."

The question of whether visual imagery is prominent in a reader's re-creative imagination was seen as a pressing issue, one Scarry was very much aware of as she wrote *Dreaming by the Book*. Richardson is forthright; "Scarry's analysis ... restores the continuity between the visual image and creative imagination that 'antipictorialist' accounts of literary imagery deny or discount" (Richardson 2010a: 38, 46, respectively). Thanks to neuroscience, as we hope to establish, this imagery debate has now been resolved beyond reasonable doubt.

Before tying Scarry's work back to its roots in experimental neuroscience, it helps to see her ideas applied in some working examples, like *The Inheritors*. With such practical application in mind, at this juncture it is useful to introduce another novel set in prehistory, *The Gift of Stones* by Jim Crace. Both novels have much to say about the vital role of storytelling, and (in the artfulness of their own construction) how it functions to spark imagination. They are, in very different ways, stories about stories.

Shadows in the Cave

Scarry confronts the key question, "By what miracle is a writer able to incite us to bring forth mental images that resemble our own perceptual acts?" And in doing so she notes that in some arts—painting, theater, film, music—vivacity of imagery is to be expected because the art itself has "immediate sensory content." But oral and written narrative is almost totally devoid of such content. When an author, or raconteur, describes someone's face, we might also say they confer a set of instructions for how to imagine, or construct, it. Put another way, instead of sensory impact, the narrative arts have "mimetic content"—they instruct our brains how to mimic or represent (see Scarry 2001: 7).

Scarry's observation shifts the focus of the traditional concept of mimesis (e.g., Erich Auerbach [1953] 2003) away from "objects" (the text, the performance) to the mental act of representation in the reader, or viewer. And where does all this lead? Well, to the loaded conclusion that the apparent reality of events described in a novel comes about by "*reproducing the deep structure of perception*" (9). This last phrase is paramount. Yet it seems many reviewers didn't appreciate its massive implications.

Scarry begins by confronting something most of us are disinclined to admit: the thinness of our imagination when compared to actual perception. The reason for this insubstantiality, she says, is that imagination is parasitic upon perception and memory, is an echo, a secondary phenomenon. The description of "a log cabin by a stream" may invoke many "log cabins" and many "streams" in various readers' minds, at different levels of vivacity and detail depending upon their life experiences, neurophysiology, and unique memories.

Only "by decoupling [the idea of] 'vividness' from 'the imaginary' ... and attaching it to its proper moorings *in perception*, can we even recognize, first, that the imagined object is not ordinarily vivid, and, second, that its not being vivid is tautologically bound up with its being imaginary" (4). There is, she points out, a "general vacuity" around images conjured up by words: they have a tendency to be flat, unrealized at the edges, tenebrous. This lurking emptiness, she suggests, is one reason for our perennial interest in ghost stories:

> Why, when the lights go out, and the storytelling begins, is the most compelling tale (most convincing, most believable) a ghost story? ... The answer is that the story instructs its hearers to create an image whose own properties are second nature to the imagination; it instructs its hearers to depict in the mind something thin, dry, filmy, two-dimensional, and without solidity. ... It is not hard to imagine a ghost successfully. What is hard is successfully to imagine an object, any object, that does *not* look like a ghost. (24)

Peter Mendelsund admits as much in *What We See When We Read*: "I am a book designer, and my livelihood depends ... on my ability to recognize the visual cues and prompts in texts. But when it comes to imagining characters, daffodils, lighthouses, or fog: I am as blind as the next person" (23). Needless to say, one of the writer's first tasks, regardless of genre, is to achieve Scarry's "hard" goal, to liberate the image from filmy evanescence and confer "solidity," vivacity, and dimensionality.

Fully achieved imaginative engagement is not mere reverie, however. Scarry makes a strong distinction between the "enfeebled condition of daydreaming" where the inner pictures exist in a kind of inchoate limbo, and the "mechanisms of vivid imagining" set in motion by the most powerful literature. The transformation of one into the other is the overriding aim of the storyteller.

In pursuit of this aim, Scarry gives many examples of writers instructing us exactly *what* we should envisage *when,* and *how* we should do so. As she stresses, the textual placement of such cues is crucial. Their location, tempo, and sequence within the narrative matter greatly, given that reading is a successive, unfolding experience like playing a musical score. Scarry points to a syntax of instruction—analogous to an IT operating program—lurking within the text, animating and enlivening the more conspicuous and overt syntax of description. In this respect, the writer resembles, to use Golding's expression, a sorcerer, or the conductor of an orchestra, invoking, inciting, and controlling our "inner vision" at every turn, with our own connivance.

The Practical Choreography of Ghosts

Jim Crace's *Gift of Stones* (1988), a story about a Paleolithic storyteller, provides evidence of many clear and, since this is a metanarrative, self-conscious examples of just the kinds of authorial prompts Scarry identifies. In the passage below, the storyteller's daughter, using many of her father's narrative techniques, imaginatively recreates an event in his childhood that shaped his life: the loss of his arm to a poisoned arrow. Authorial "prompts"—hidden or merely implied—are extrapolated and rendered in italics by us.

> Here, perhaps, a picture of my father as a boy should take its place between the bracken and the riders and the sea. [*I am asking you to set the scene now in your mind, triangulating the boy's position using the elements I give you*] He was in his seventh year and though there were children of his own age and younger whose weight and muscles had matured, he was still a bullrush of a boy, a stem, [*visualize the thinness of his body in "long-shot" alongside the image of a bullrush, specifically its stem*] his elbows—both elbows still—[*remember, this image of completeness I am giving you is only temporary, as he is going to lose one arm*] thinner than his arms, his chest as flat and formless as a slate. [*now "pan" across these bodily features, and juxtapose the image of a slate*] His cousins said his face was disobedient and dreamy [*summon up such facial expressions from your memory*] a combination which they found more than doubly irritating. [*try to make a composite of the two facial memory impressions*] Perhaps it was this challenge and this indifference in his face [*now superimpose a look of indifference onto the new composite*] which caused the riders to treat him roughly. They paced their horses round him and one put out his hand to take the scallops [he carried]. [*"zoom out" now to see the size of the riders relative to the young boy*] My father was small and fast and unafraid of horses. He rolled beneath a mare [*animate the image of the boy, rolling*] and disappeared into the bracken. [*now, lose the boy entirely for a moment*] And then he showed himself again, [*reinstate his image, suddenly*] standing and jeering on a rock where the horses could not reach.
>
> The picture is incomplete. What he did not see, what only now I can construct in my imagination [*all the foregoing, everything you've*

imagined thus far, has actually been a story related to me by my father]
and place a little distance from the horses, [*so, I'm now the one dictating
the scene from my own imagination*] was one other man, dismounted,
bow raised and drawn, arrow loosed. It struck the boy, my father, in the
arm below the elbow in the arc of flesh which hangs like cobwebs from
the bone [*see this impact in extreme close-up in conjunction with the
image of a cobweb, its sense of fragility*]. (3–4)

These authorial cues (explicit and implicit) are there to summon up our
imaginative faculties using something already latent within us. And that
"something" is memory.

To use the more inclusive term from psychologist Frederic Bartlett ([1932]
1995), imagination is formed from a "schema"—taking us back to Richards
again—meaning the sum of everything we know, our mental models of the
world, be they derived from memory, dream, or indeed the written word.
"The term is based on a schematic drawing, for instance from an instruction
manual ... our schemas develop continually, and are internally coherent:
working models of what we know about how the world works" (Oatley
2011: 61, and see Chapters 6 and 9).

In the passage from *Gift of Stones*, there is a sudden recursive dislocation
marked by the break of paragraphs: "The picture is incomplete," which
reminds us of one of the chief tools available to the writer as *auteur* of our
inner schema—shift in point of view. Here, Crace violently pulls the rug from
under our feet, by revealing that this straightforward story of a boy in trouble
(level 1) is in fact a story within a story (level 2) and that moreover, level 1
may well be an untrustworthy account—the word "perhaps" recurrent—
and that the author of level 2—the storyteller's daughter—is having to fill in
the blanks by way of her own guesswork ("he did not see, what only now
I can construct in my imagination") prompted by her memory of hearing
level 1. Not only are the accounts nested then—her account a retelling of
his—but her current one which seems definitive (level 2) actually involves
considerable (imaginative) conjecture.

Our sense of what is "authoritative" is jeopardized. The daughter,
having heard her father relate this anecdote *ad nauseam*, is "re-envisioning,"
"re-presenting" it for us in her own way, and in the process deviating from,
and distorting her father's authorial instructions. Crace has fun here with
the precariousness of our trust in the "inner vision" provoked by story and
delights in showing us the false bottom in the magician's cabinet of dreams.

Scarry has much to say about related narrative skills. "It is difficult
to move or to animate the image ... [it] cannot be easily turned upside
down ... and the image (of a face, for example) inhabits a generalized
posture and cannot be shifted through an array of angles as would occur in
perceptual reality" (Scarry 2001: 35). The implication is that if an author
can somehow persuade us into undertaking some of these "difficult"
maneuvers ourselves, it will go a long way to convince us that the scene in
our minds has veracity.

She enumerates many strategies for such actualization, each devised to counter the "tissuelike" or "flimsy" quality of the mental image. One of the most basic is to ask the reader to watch the image move: or, more precisely, entice them into moving the image themselves. So, in the passage above we have, "They paced their horses round him and one put out his hand" and "He rolled beneath a mare," both phrases demanding that the image is given heft, trajectory, momentum: that it shift from the photographic to the cinematic.

In a chapter entitled "Stretching, Folding and Tilting," Scarry takes us through some other ploys used by writers that convince us of the tangibility and vibrancy of the scenes they summon up; again, starting from her belief in "the tissuelike quality of the image, which seems to have the transparent thinness of a film of skin on which a picture has been projected."

Scarry asks how this tenuousness can be countered. The answer is to confer on the image the qualities of objects in three-dimensional space. Anything which makes the simulacrum seem tangible brings it that much closer to perceptual reality, and since we know that mental picturing uses many of the same neural processes as sensory perception (see later chapters) this makes good sense. So when Crace writes "He rolled beneath a mare," our memory of seeing a body curled up and tumbling (in this case beneath the belly of a horse) enhances our sense of a figure moving in a recognizable way through space. It breathes life into what could otherwise be a flat, undifferentiated, and static image of a young boy standing.

If animating the image is demanding for the reader, there is another strategy of actualization that is less so. It is what Scarry calls "addition and subtraction," a technique that cleverly exploits the ephemerality of the image to its own advantage.

> If an image is present, then disappears, it seems to have moved. This act of subtraction is an easy operation to perform, since it takes almost as much mental labor to sustain an image over three, five, or twenty-five seconds as to compose it in the first place. Permitting it to vanish requires no work, since left to itself the image vanishes on its own. (100)

So indispensable is movement to the composition of vivid mental images that even the approximation of it, created by the simple children's game of "now-you-see-it-now you don't," heightens our sense of the real. As with cinema which traditionally used twenty-four (static) frames per second, the technique of rapid subtraction and addition uses the limitations in our conscious awareness to simulate movement.

Jim Crace, because he is composing a story about storytelling, is thus liberated to exploit every possible authorial ruse. Hence, the boy rolled under the horse "and disappeared into the bracken. And then he showed himself again, standing and jeering on a rock where the horses could not reach." A classic "peek-a-boo" case of subtraction followed swiftly by reinstatement, you might think, but he goes one step further, speaking in the voice of the daughter, "What he did not see, what *only now I can construct in my imagination and place a little distance from the horses,* was one other man, dismounted, bow raised and drawn, arrow loosed."

What is beautifully described at this point is imagination "caught in the act": the daughter very deliberately choosing where she wants to "place" the attacking figure in (imaginary) space, "a little distance from the horses." By doing this, she undertakes an "addition" of her own to her father's story, instructing us to suddenly see what he could not: the dismounted rider and his "loosed" arrow.

As the novel progresses, we realize it is set in a village community whose members survive by mining and knapping (shaping) flints. The protagonist's physical handicap, once he loses his arm, means he can never contribute to the work of the village, so his promotion of imaginary worlds by way of storytelling acts as some sort of compensation. The storyteller is both less and more than a man.

After the trauma of the amputation, the storyteller struggles to find a role for himself. So when he sees something highly unusual and alarming on one of his wanderings—the arrival of a ship just offshore—he initially ignores it, then, when he finally decides to tell the tale, he can't resist embellishing. "We know that when he spoke he shaped the truth, he trimmed, he stretched, he decorated. He was to truth what every stoney was to untouched flint, a fashioner, a god" (58). It is no coincidence that trimming, stretching, and decorating are among the key narrative techniques identified by Scarry. Sure enough they have their desired effect for the prehistoric storyteller. He "could now hold a household silent with the magic of his words." Entirely through the power of narrative, he has undergone an apotheosis from cripple into bard, and even, "god."

"He could not, he said, have invented a more workable device for telling tales than the ship upon the sea. Each time it came ashore it could offload a new and untried plot; a different set of characters" (59). By great good fortune, his serendipitous discovery contains all the factors—movement, arrival, disembarkation, disappearance, and reappearance—to activate Scarry's fictional "sleights of mind."

As David Herman succinctly puts it, "Some stories, in addition to enabling interpreters to project a storyworld, can also foreground, through their very structure, a concern with the nature and effects of such world-projecting acts" (2013: 68). Crace's storyteller, through his narrative embellishments and artistry, entrances his village, but in the process ends up losing sight of his own truth, and becomes the bearer of ill-tidings—the arrival of the ship—which will mark the end of their way of life.

Entropy and Extinction

The Inheritors also centers on a period of violent prehistoric transition. Golding's novel presents us with an opportunity to explore a working hypothesis not only about the role of storytelling in our development, but about the evolution of language itself. This tale is fascinating because it ventures way beyond the social aspect of narrative into the earliest glimmerings of human thought. Yet even in this strange experimental work, which is obliged to invent its own protolanguage, many of Scarry's insights on animating and vivifying the image hold true.

Golding depicts evolution in action by depicting the moment when *Homo sapiens* first encounter Neanderthals (see Chapter 5) but elects to narrate the tale mainly from the point of view of the latter. The Neanderthals are beings with acute hearing and smell, but who struggle to connect their thoughts. For them everything—the streams, the sky, the forest, and the land—is pantheistically alive and interwoven, and they display a largely visual-pictorial way of thinking, more compelling than their rudimentary language.

All of this makes the narration of their tale problematic for both author and reader. The barriers to comprehension are even higher because most of the story is filtered through Lok, one of the least insightful members of the Neanderthal tribe. And that, at least partly, is the point. By obliging us to extend our own mindreading skills to these creatures, he is pointing up the habitual limitations on our own empathy, and asking of us who we care about and why. Ultimately, the tragedy of the encounter between the two groups is a direct consequence of our human ability to withdraw empathy, something impossible for the Neanderthals. Seen in this light, what Golding presents us with in *The Inheritors* is little less than a cognitive "experiment" in the limitations of mindreading.

What's telling is that, despite the boldness of his attempt to depict minds quite foreign to our own, Golding as novelist is obliged to rely on many of the same fictional strategies identified by Scarry. Stretching, folding, tilting, addition, and subtraction of the mental image are all seen frequently in the novel. It is a reminder that, however far fictional travel takes us (even, as here, down an evolutionary *cul-de-sac*), we are still constrained by the neurological and cognitive laws that comprise our own genetic heritage—our own Umwelt. For most of us, most of the time, our imaginative experience acquires color, dimensionality, and life due to the deployment of just such narrative contrivances.

One of the most important of such devices Scarry calls is "radiant ignition." Citing Homer's *Iliad*, she demonstrates that the technique has an ancient lineage. But what does the term mean? In a thought experiment, she asks us, first, to imagine a dark featureless space, then to introduce a sudden pinprick of light, finally making it streak across the blackness, as if it were moving across the retina: "Imagine this dark space with sudden flares and lights—bursting, then disappearing, streaming across the field of mental vision or arcing through" (Scarry 2001: 77). This, she says, is exactly what Homer asks us to do in his epic poem. "Homer gets us to imagine motion by producing a steady surge of ignitions. It is not just vast armies and random arrays of flickering lights, but highly individualized, discrete motions that he requires us to see" (78).

And here is Golding, a scholar of Greek, describing his protagonist, Lok, at first sitting by the fireside, then, feeling a certain unease, staring off into the surrounding night:

> He looked up at the sky and saw that it was clear except where layers of fleecy cloud lay above the sea. As he watched, and the after-image of the fire faded he saw a star prick open. Then there were others, a scatter, fields of quivering lights, from horizon to horizon. His eyes considered the stars without blinking, while his nose searched for the hyenas and told him that they were nowhere near. (Golding 1988: 40)

We are given the expected eidetic triggers, the "layers of fleecy clouds" and "fields of quivering lights." But he also prompts us to inhabit the mind of Lok in a surprising way, "As he watched, and the after-image of the fire faded he saw a star prick open." Golding here is not simply offering us a fleeting visual impression, but evoking a physiological fact: the existence of after-images on the retina. By alluding to this universal experience of image "after-burn," he emphasizes our commonality with this naked, bemused creature. At a stroke, he forges an intimacy between the reader and Lok.

Indeed, the passage subtly suggests that it is not only the lights which scatter, but Lok's very being which is subject to disintegration, fragmentation. Hence it is his eyes that "considered the stars" and his nose which "searched for the hyenas," rather than Lok himself. A sense of dispersed agency is created. This reinforces our impression of him as one aspect of a larger whole, and also presages the catastrophe that is to befall him and his kin. There is an aching vulnerability here. The passage continues:

> There was always light where the river fell into its basin. The smoky spray seemed to trap whatever light there was and to dispense it subtly. Yet this light illumined nothing but the spray so that the island was total darkness. Lok gazed without thought at the black trees and rocks that loomed through the dull whiteness. (40)

This is a compound example of both the manipulation of light and the use of translucency to emphasize dimensionality. Scarry states early in her book that "the passing of a filmy surface over another (by comparison, dense) surface" is a key way of conferring solidity on the image, which, we recall, has a tendency to "ghostly" immateriality in our mind (14). By showing us the "black trees and rocks" "through the dull whiteness" of the spray, the features of the island become more substantial.

On close inspection, what Golding is doing here is an attempt at the recreation of Lok's own perceptual uncertainties, his cognitive stutterings and revisions, in the mind of the reader. Thus, each time, as readers, we approach imaginative clarity in the description, the writer trips us up, undermines our assumptions. Any facts ("There was always light") are immediately subverted "Yet [it] illumined nothing." The result is a perceptual seesaw which is deeply unsettling. In this section of the novel, we are repeatedly asked to "see" something in the mind's eye, and then instructed to "unsee" it again; or else we are asked to visualize what cannot be seen (the equivalent of Milton's "darkness visible," the title of a later Golding novel). The result is cognitive dissonance of a dynamic and unstable nature. The writer sets us up to experience the restless ebb and flow of Lok's incomprehension for ourselves, and one of the ways he achieves this is to problematize the mimetic act itself.

Elsewhere, in essays and interviews, it is evident that Golding retains a very practical grasp on the essentials of his craft:

> We ... [have] each of us a separate awareness in the circle of the skull, each willing to have what the author offers inside the circle ... He must, like a sorcerer, raise up appearances, simulacra in the circle of the skull, which will take on the likeness of individual men and women in whom we may become deeply interested, and whose strictly unreal, paper fate may inundate us with amusement, curiosity, joy, grief. (Golding 1982: 137)

This is the novelist as magus, but one who needs always to exercise fine judgment to maintain the illusion. Magicians are preoccupied with viewing angles for good reason. Those "simulacra in the circle of the skull" need to be carefully focalized, sharpened, and intensified. The "ghostly" inner images must acquire manifest reality. As author, he is clear that this is his primary job. If fiction can be compared to a map, it is not like a GPS or a satellite positioning system, but rather a PPS, or personal positioning system—where the person is not me.

The Neural Frame

In what ways do the theories of Elaine Scarry intersect not just with acts of cognition (mind), but also with their actual substrates (brain)? There are at least two possibilities. The first has to do with the formal connection of her arguments to brain science, and provides an intriguing answer for why Crace chose the working of flint as his central metaphor.

All the authorial strategies Scarry outlines in her study have a single goal: to make the phantom image appear tangible, or, in her words, "handle-able." She turns to neuroscientific findings in developing a theory that turns out to be highly applicable to the flint-knappers of *Gift of Stones*:

> A map of the relative size of body parts as they exist not in the body but in neural activity [see Chapter 4] shows by far the largest body part is the hand ... it is as large as many other parts combined .... It may be that

alluding to the hand brings large resources of the brain to bear on the project of making an image move. (Scarry: 147)

Even more significant, in terms of the Crace example, Scarry notes:

Recent work on mental imaging in cognitive neuropsychology has shown that the part of the brain at work when one thinks of a handmade object (a chisel, a doll, a house) is not the same as when one thinks about a natural object (a stone, a shell). It also turns out that the part of the brain engaged in thinking about handmade objects is the region engaged in thinking about motion. (146)[4]

So, by portraying a prehistoric craftsman and his kin engaged in the act of transforming "non-handmade objects" (stones) into "handmade objects" (e.g., blades), Crace may be inducing us to cognitively animate his creations, to give life to otherwise inert tableaux, by activating new parts of the brain.

As the Bronze Age storyteller tells, re-tells, and embellishes the pivotal tale of the arrow that cost him his arm, he goads the audience with his extravagant virtuosity:

It was many years ago and I have told this story many times and changed it just as often. But one thing never changed. The bowman's face, his smile, his eyes, expressed in full what neighbours in our village had most distrusted in my own face. Look, you see it now, a little blunted, true ...So this is my story then? Watch out, you say he's chipping and he's knapping at the truth. He's shaping it to make a tale. (33–4)

Often, as here, there is a tacit admission in the novel of something unnerving, even dangerous about the creation of stories. They induce "dreams," "turbulence" (34) and, unlike the straightforwardly physical "chipping and ... knapping" of stone, can lead to an undermining of certainties, even of the self. When the raw material is as corruptible as memory, the very act of retelling can blur the boundaries between the real and the imaginary, our ability to "tell" the one from the other—even for the teller: "I can't be sure. It was many years ago and I have told this story many times and changed it just as often." As we will see (Chapter 9), the stuff of memory is volatile and, even without artistic license, has a tendency to warp, corrode, and refashion itself.

The second way Elaine Scarry's ideas connect to brain science is through the bedrock assumption behind her theories: the conviction that reading words generates images in the mind—an idea clearly manifest in I. A. Richards model. Scarry was certainly familiar with findings from twentieth-century cognitive neuroscience about the generation of mental images and cites them in her book, first published in 1999. At that time, functional brain imaging was just beginning to yield relevant results.

It turns out when we imagine objects or locations with eyes closed or in the dark, there is accompanying neural activity in the visual area of our cerebral cortex, just as there would be if we had actually seen them. Even more compelling, when experimenters alter the neural processing in these cortical areas, there are corresponding changes in what people report about their imagined "scenes" (Slotnick et al. 2005, Stokes et al. 2009, Albers et al. 2013). Just as with the "inner voice," the weight of evidence has led to a broad consensus that the "imagery debate" has been resolved. This has large implications not only for understanding our mental lives generally, but also for verifying the role of images in reading stories.

The imagery debate is about whether we represent information in our brain only in language-like (propositional) terms, or also pictorially. After fifteen years of increasingly clear experiments, Kosslyn and colleagues ended the debate: we do not represent fictional events only in words, but can also represent them graphically. When we read a fictional description, neural activity in the visual areas of cerebral cortex is similar to the activity there when we acquire images during perception. So to see or imagine "with the mind's eye" is, after all, qualitatively like visual perception (Pearson and Kosslyn 2015, Pearson 2019). The specific notion of "seeing" when we read (see Richards, Mendelsund, and many others) is backed up by recent functional brain scans (e.g., Brosch 2018); moreover, the scans can detect differences in vividness and confirm that imagined images are typically less vivid, as Scarry claimed, than those of direct perception (e.g., Dijkstra et al. 2017). We will look at the links between Scarry's techniques for orchestrating the imagination and recent ideas about the centrality of imagination in the origin of human language, in Chapter 5.

To return to Heaney, who, prompted by Thomas Hardy's *Return of the Native*, was so disoriented that he found himself unable to tell the surface of a field from the surface of the page, in the light of the above, it should not surprise us that it was this particular novel which induced such a blurring of sensory maps. Hardy, we know, spent the year before writing it researching the geographic location which he would take as his fictional setting of "Egdon Heath": walking its paths, making sketches, charting its features, researching its flora and fauna, before describing it in such detail in the early chapters that his publisher complained that the plot was much delayed. But Hardy knew what he was doing. He sensed that imagination, let alone transport, derives from perception and memory working in harness, and, finally, depends on the obdurate dimensionality of the real.

Notes

1   In clinical neuroscience it was believed that "neurasthenia" was a common condition (diagnosed especially in women and those of artistic temperament)

for over a hundred years, until it was discredited. There are numerous other examples.

2   See Rimmon-Kenan (2002), for concise analysis of the shifts between an "external and internal focalizer," and (2015) on how narration provides special access to subjectivity.

3   James Wood's *How Fiction Works* (2008) offers a lucid description of the effects of free indirect style. Based on it, "We inhabit omniscience and partiality at once. A gap opens between author and character, and the bridge—which is free indirect style itself—between them simultaneously closes that gap and draws attention to its distance" (10). For early analysis of speech acts, see Austin (1962) and Searle (1969).

4   This quotation refers to work described in Hanna Damasio et al. (1996); and Alex Martin et al. (1996); reference to brain activation during imagined motion should be to J. Decety (1994).

PART II

Into the Neural Terrain

In 1919, Virginia Woolf hailed the value of modern writers like Joyce who enabled a reader to "imagine what he can neither touch nor see" by revealing "the flickerings of that innermost flame which flashes its messages through the brain" (151). In these two chapters, we describe the neuronal circuits that support the cognitive networks of our brain, allowing an "encultured" framework for representing the world and other minds.

3

Brain and Behavior

In a book review in the *New York Times* (August 19, 2012), William Giraldi revealed, "Arthur Miller's original title for *Death of a Salesman* was 'The Inside of His Head,' until he came to his senses: all of literature is about the inside of somebody's head." The literary arts have long been recognized as profound sources of insight into our inner life. In a different but related way, psychology and neuroscience have a similar interest in "the inside of somebody's head." What happens there is what most makes us human. So brain scientists can't help finding literature an attractive focus for their efforts, recognizing that it often derives from the same motivation that drives them. Neurobiologists follow this up directly by looking inside the head to observe meaningful correlations of brain physiology with behavior. Cognitive scientists do it indirectly by defining algorithms that can explain or mimic mental processes.

We hear claims every few years about the death of the author, the critic, literary culture, the novel, or drama. Paradoxically, these claims often come from within the literary world.[1] However, from the perspective of neuroscience, there is much evidence that literary practices are sufficiently grounded in our brain function and psychology that they are unlikely to suffer extinction any time soon. Even as the once-dominant medium of words-on-a-printed-page diminishes, narrative art will simply adopt other forms. As Mark Turner (1996) put it: "Narrative imagining—story—is the fundamental instrument of thought. Rational capacities depend upon it. It is our chief means of looking into the future, of predicting, of planning, and of explaining. It is a literary capacity indispensable to human cognition generally … *the mind is essentially literary*" (4, our italics).

To approach stories and literature from a behavioral and neurobiological perspective is not to replace the traditional analysis of narratives and their capacity to convey the emotional richness of human life. Shakespeare and Proust, Woolf and Beckett, Byatt and Atwood, McEwan and Doctorow, and myriad others map the human psyche in ways that are certainly as important as those charted with the objectivity and thoroughness toward which science strives.

If our passion for stories and literature is somehow a part of human nature, our brain is not just one place to look—as the "organ of behavior," it's the essential place to look. We will begin by describing a human brain in its bodily and environmental contexts, and then see what its fabric, its cells and their interweaving circuits, can tell us about cognition.

Intellectual Antecedents of Neuroscience

Compared to the long and distinctive lineage of literary criticism, neuroscience wasn't really a separate, unified discipline until the latter half of the twentieth century. However, its modern roots extend back to classic behavioral observations by Charles Bell, Charles Darwin, and William James, and research by cell biologists, physiologists, and anatomists such as Camillo Golgi, Santiago Ramon y Cajal, and Charles Sherrington. William James' *Principles of Psychology* ([1890] 1950) synthesized mechanistic behavioral science and physiology just before the twentieth century began— and it is often forgotten that Sigmund Freud was a neurophysiologist who worked in several areas, including the neurological basis of language, well before he became intrigued by stories and dreams, culminating in the development of psychoanalysis.

Because neuroscience is in part a biological discipline—with evolution by means of natural selection as a core organizational principle—it displays several attributes derived from evolutionary theory. Like the rest of biology, neuroscience is intrinsically historical. It is concerned not only with traits or functions as currently observed (such as *how* behaviors are produced and controlled), but also with origins and design principles (*why* particular behaviors emerged and are organized as currently observed).

A second and closely related point: brain function and behavior are understood as the accumulated results of survival forces operating over time *without any planning or design*. Evolution is not a directed process. Furthermore, the possibilities for trait changes at any point in time show *contingency*; they are inevitably constrained by what happened earlier. So for example, the brains of mammals like us emerged from progressive alterations to a basic "bauplan," or organizational map, of the vertebrate nervous system, not from being rebuilt with new materials.

Finally, biology brings to neuroscience what Ernst Mayr called "population thinking." That is to say, individuals do not evolve, rather it is populations, or species, that do. This notion was central to Darwin's thinking: within any generation there are slight differences from one individual to the next within a population, that is, *variation*. Some individuals will be better able to survive, so their descendants will tend to be more frequently represented in subsequent generations. This is *natural selection*.

In Darwin's time, the genetic basis of variability in traits and their inheritance was unknown, but we now appreciate that variation is typically driven first by random processes, often mutation, altering some of an individual's genes, and that evolution then changes the frequency of genes within populations over time, based on this selection. Such variability is a prerequisite not only for natural selection but also for *sexual selection*. In natural selection, the factors leading to differential reproduction are environmental; in sexual selection, differential reproduction depends upon intra-species *behavioral* interactions, such as male rivalry and female choice.

The experimental mindset of life scientists may cause them to see differences between individuals of a species as a distraction, perhaps even an irrelevance, on the way toward a clear-cut experimental result, so they are sometimes ignored. However, such variation of traits is often the most important part of what is studied in neural and behavioral work, and a source of great insight.[2]

Precisely because variation is ubiquitous in living systems, it is important to be aware that *essentialism* is a mindset that can ensnare scientists *and* nonscientists, resulting in stereotypes based on unfounded generalizations. Essentialism assumes that we can "carve nature at the joints" and identify groups (species, races, etc.) based on universally shared, unchanging essences. Any biologically informed discussion that falls into this trap and ignores the profound variation between individuals is, quite simply, bad science.

The Ecosystem of Mental Life

Humans have speculated about the workings of mind ever since we acquired our capacity for self-reflection. As Colin Blakemore (1977) has noted, "The debate that raged, since at least the time that men [*sic*] first wrote down those things that troubled them, was not about *whether* the mind had a physical counterpart in the body, but *where* its embodiment might be" (9).[3]

There were several candidates. Historically, the brain's main competitor as the animator of life was the heart. This was based on common sense: it is always active during our lives; its pattern of beats changes as our levels of arousal and activity change. The idea that our brains host our mental experiences seemed outlandish before we collected observations and tested possibilities with something more systematic than common sense.[4]

The brain's very appearance is somewhat challenging in this respect. It weighs about three pounds and, in its native state, has a consistency only slightly more solid than rice pudding. It is held in its characteristic shape by the vault of the skull and a tightly enclosing system of membranes— the *meninges*. In contrast to the dynamically pulsating heart, brains seem to have no moving parts and a lumpy appearance that is unremarkable. However, careful anatomical analyses over several centuries, and at ever finer levels of detail, provided a clarifying context: our nervous system displays tremendous dynamic activity, and *has* moving parts, but these facts are only discernible at the cellular and molecular levels.

Our brain also includes a complex network of arterial and venous plumbing. Thus early in history, the brain was sometimes thought to perform the humble function of cooling the blood. What turns out to be more important is what the blood does for the brain: it provides oxygen and chemical energy (blood sugar or glucose) and acts as a conduit of communication between brain and the endocrine glands and the immune system components distributed around the body. These three control systems are in continuous dialogue.

It's notable that blood flow to the brain is finely regulated from second to second because of constant need for glucose by active brain cells. Modern functional brain scans typically tap into blood flow, non-invasively, to detect when the activity of brain cells has changed. While the real technological development of the two most common forms (positron emission tomography [PET] and functional magnetic resonance imaging [fMRI]) began in the 1940s, it was actually Charles Sherrington who made the key observation enabling these developments. Work in his lab showed that blood flow to local areas within the brain varied along with the metabolic needs of those areas (Roy and Sherrington 1890).

So a region with more active neurons will selectively receive increased circulation: PET detects this via regional changes in the levels of harmless isotopes injected into the blood; fMRI requires no injections and detects changes in blood's natural oxygen carrier, hemoglobin, through a magnetic signature of sorts. Because their activity is related to communication, we will shortly describe the electrical signaling of brain cells in more depth. As we do so, be aware that recordings with microelectrodes can detect these signals directly, PET and fMRI do so indirectly: changes in blood flow are used to infer changes in neural activity.

Body, Brain, and Mind

Many of the facts cited above influence how we think about our brain in broad terms. Scientists often place the brain in context using the adjective "embodied." As a principle of neuroscience, the term conveys the fact that the nervous system interacts continuously with the endocrine and immune systems, so our brains are immersed in a circulating sea of bodily information, with their own tides and flows of hormones, metabolic factors, and pathogens. Embodiment also denotes that sensation from the body itself—about posture and movement, as well as the state of the viscera—is part of the core input that shapes our perceptions and behavior, even if unconsciously.

There is actually a long history to this idea, given that, during episodes of emotion, when we have experiences such as fear, anger, joy, etc., there are easily measurable changes in the body—by way of the autonomic division of the nervous system. Heart rate, blood flow, and respiration are altered;

glands are activated affecting energy production and water balance, to prepare for possible action. As a consequence, emotions are not just mental, they are also corporeal (Chapter 7). Any model of a disembodied brain with a one-way causative arrow pointing toward behavior or cognitive states would be "non-ecological" and anachronistic. More accurate is the depiction of our brain with multiple, shifting two-way information flows between it and the other regulatory systems of the body, with other brains, and with culture (Figure 3.1).

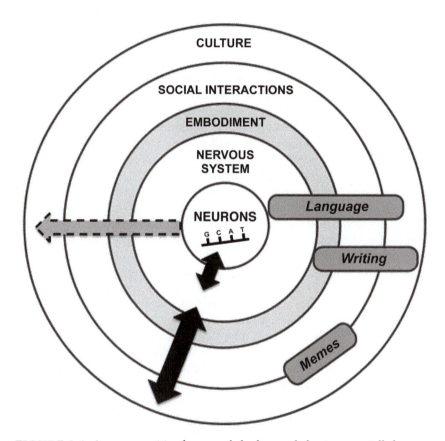

FIGURE 3.1 *A neurocognitive framework for human behavior, especially language and literature. Central circle represents the basic structural and functional units of the nervous system, neurons, and the inherited genetic information of an individual that they contain. The next ring is the full nervous system and the light gray ring represents systems that literally embody the nervous system (muscles and mechanical constraints, glands, immune tissues). One-way arrow (left, dashed) is a limited view of brain as producer of behavior. Two-way arrows below (black) show genetics and brain function influencing,* and *being influenced by, bodily conditions, other individuals, and culture. The gray boxes locate (very approximately) the level of influence for some language- and literature-based phenomena.*

Neuroscientists usually acknowledge this. Consider the wonderfully succinct statement from Michael O'Shea (2006), "The word 'brain' is a shorthand for all of the independent, interactive processes of a complex dynamical system consisting of the brain, the body and the outside world" (3). As a result, the "ecology of mind," to borrow a term from Gregory Bateson (2000), includes environments both physical and social, but it remains centered squarely on brain physiology. Such a dynamic framework is indispensable to bringing stories, literature, and brains into the scientific understanding of mental life. The two-way arrows indicate everything that not only our brain influences, but also the many return influences on our brain: the body, social networks, and also culture at large (including stories and literature).

Beyond the consideration of emotions, there are recent theories that propose cognition generally, or specific domains like language, may be embodied. For example, perhaps our very understanding of words and concepts depends not only upon higher-level, abstract symbol processing, but also on activation of sensory or motor circuits, where experiences and actions may be simulated. We will unpack some implications of this extended sense of embodiment in more detail in chapters ahead.

As Figure 3.1 suggests, our brain can be described as the central node in a larger ecosystem that supports the many dynamic interactions of mental life, though even this does not do it full justice. Human brains are complex and heterogeneous organs, containing 80–90 billion neurons, and a larger number of connections among them by several orders of magnitude, together with similar numbers of glial cells. If all of the potential interactions between the components of Figure 3.1 could be mapped at anything like that level of detail, there would be arrows passing in all directions across every level and the diagram would begin to resemble the complex charts ecologists use to represent flows of energy and information within entire ecosystems, only more so. To place our brain in its bodily "habitat," we start at the systems level and then look internally to its cells, before moving back to consideration of more cognitive and cultural understandings of language, listening, reading, and stories.

Of course, the brain is just part of a larger entity. Together with the spinal cord it comprises the central nervous system (CNS), plus peripheral nerve pathways connect the CNS to input from sensors (eyes, ears, etc.), and motor pathways leading to effectors (muscles and glands). Also, there are numerous types of sensors on the input side: *exteroceptors* respond to energy impinging on the body (e.g., light, sound waves, touch, etc.), *proprioceptors* report on self-action (via muscle and joint receptors), and *interoceptors* monitor the state of our viscera (e.g., blood pressure, respiration, gut function).

The variety of receptors, and their respective levels of penetration into conscious awareness, explains some of the complexity and subjectivity of each person's sensory world. On the output side, there are a corresponding number of action channels accounting for behavior, overt and covert.

We have just described sensory and motor pathways in the way people often think of them: sensors sending messages toward the CNS, and the CNS sending signals toward muscles. However, consistent with our use of two-way arrows above, it is true that there is actually two-way traffic in most sensory and motor pathways, and between levels of the CNS. For instance, as sensory signals ascend toward higher centers, so-called bottom-up processing, those pathways also have some axons (see below) carrying descending signals that can regulate sensation and perception, "top-down" processing. Mental phenomena emerge from a combination of both.

"Michelin Guide" to the Brain

Some familiarity with brain structure, the physiology of neurons, and the "circuit thinking" in which neuroscientists engage will help in negotiating the information in the next few chapters. This brief overview can be skipped by anyone already conversant with neuroanatomy or basic neurophysiology (for details of brain structure and evolution, see Butler and Hodos 1996, Striedter 2005).

Navigating vertebrate brain structure is easier if you know that it consists of a small number of component regions and their linkages, just as navigating New York is somewhat easier if you know that it is composed of five boroughs. As indicated, our brains conform to an organizational scheme, or outline map, characteristic of our taxonomic group, the vertebrate bauplan. So it is relatively easy to recognize general similarities we share with other vertebrates whether fish, amphibians, reptiles, birds, or other mammals. An important caveat: this sequence is not a hierarchy. The brains of species in a class are not in any simple way descended from one another; each has a separate adaptive lineage and similarities reflect the contingencies of common developmental programs, with the particularities of species being "variations on the theme."

Figure 3.2 shows the vertebrate bauplan, based on embryology. The nervous system starts as a sheet of cells forming a hollow tube (Figure 3.2 Left). Early in development that tube begins to differentiate into four components: the spinal cord and three expanded segments toward the *rostral* (anterior) end. These rostral segments will become areas of the adult brain: the forebrain, the midbrain, and the hindbrain. Figure. 3.2 Right shows just a few notable structures that eventually develop within these basic regions. The forebrain will include the cerebral cortex and just beneath it the thalamus and hypothalamus. The midbrain will include, as its "roof," the tectum, a multisensory integration zone in all vertebrates, called the colliculus in mammals. Moving backward, the hindbrain encompasses the pons, the medulla oblongata, and the cerebellum. The general term "brainstem" is commonly used to include everything *caudal* (posterior) to the forebrain, but not the cerebellum. Many other structures are skipped for simplicity.

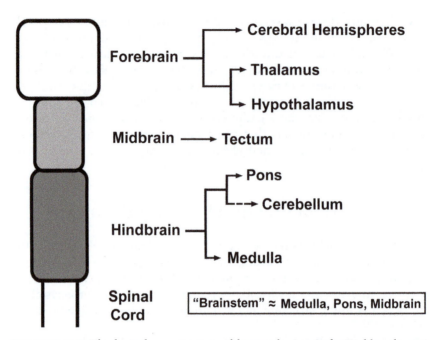

FIGURE 3.2 *The logical organization of human brains.* **Left:** *Highly schematic drawing of human "neuraxis," rostral is to the top.* **Right:** *Some key structures found within each of the three basic brain compartments. This basic "bauplan" describes all vertebrates but components may differ in relative size and structural details, reflecting the divergent history and distinct life styles of a species.*

When considering language and cognition, most references will be to regions of the cerebral hemispheres (Figure 3.3). In this figure, you can only see their heavily folded surface, known as cerebral cortex with its four lobes. Importantly, though, this diagram also shows the "primary" cortical zones for three sensory modalities that will be considered in much of what is to come (visual, auditory, and somatic). Some other structural areas of note—especially related to emotion and memory—will be covered in Chapters 7 and 9.

"Primary sensory cortex" means the cortical zone where fibers carrying information from sensory receptors first arrive at the cortical level (after being relayed through brainstem and/or other subcortical centers). Notice how small these zones are. This is because a vast amount of processing occurs "beyond" primary cortex where information subsequently flows to secondary, tertiary, and higher levels, the so-called association regions. This is where multisensory and more abstract representations occur, with individual neurons often responding to two or more sensory modalities (e.g., Stein and Meredith 1993, Sereno and Huang 2014) and sometimes to complex features such as the configuration of a face, or the emotional tone of a voice.

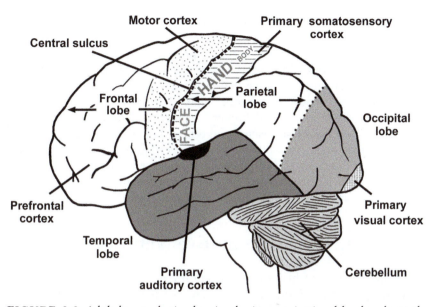

FIGURE 3.3 *Adult human brain showing basic organizational landmarks at the cortical level. This shows the lateral surface of the left hemisphere, anterior is to the left. The only detail given for a sensory area is the "somatotopic" map of the body within somatosensory cortex, where the size of each region named is roughly proportional to the amount of neural tissue representing it.*

Findings amassed over decades indicate that mammalian cortex and tectum have organizational features related to the "lifestyle" of a species. In particular, the responses of cortical neurons to sensory stimulation reveal "maps" of the body surface, reflecting the behavioral propensities of animals: for example, animals with whiskers or other sensitive appendages have detailed representations of these mechanoreceptors, and primates like us have a large amount of cortical space devoted to head sensation (face, lips, tongue) and our hands, with relatively smaller representations of the remainder of the body. This is indicative of our vital reliance on speech and tool use (e.g., Penfield and Rasmussen 1950, Suga 1989, Catania and Kaas 1997, Crish et al. 2003, Brecht 2007).

When considering maps or representations in any brain—and they are widespread in both sensory and motor regions—it's essential to be aware that the term "map," while intuitive in many cases, also has limitations. First, maps are really metaphors, they are not the terrain they represent (see Turchi 2004); second, maps are not "read" by some executive brain center. There is no equivalent of what cartographers term a percipient, noting its scale, using its codes or "legends." We take onboard Jackendoff's (2002) warning that use of the term "map" tends to imply—dangerously—an array viewed by a "little person," or homunculus, in the brain. He recommends the term "cognitive structure" instead. However, sometimes "map" is still the most easily recognized term.

Map-like structures arise because processing sensory and motor information in topographic arrays is efficient for enabling certain types of processing interactions to occur between adjacent elements. And beyond such topographic sensory or motor "maps," there are other systematic arrays that are computational "maps" representing derived information about complex stimulus properties, and others that are best called cognitive "maps" or structures. You will see examples ahead. It is also important to be clear that our brains use information from such neural arrays without any holistic synthesis—that is, there needn't be any scanning of a "map" as we might scan a road map as we to use it.

Brains as Cognitive Networks

Our understanding of human neuroanatomy is in the midst of dramatic transformation right now. Revisions are occurring because of rapidly advancing techniques, especially functional imaging. It is now possible to chart the extent of brain areas engaged during specific cognitive processes to ever more precise locations in the cerebral cortex. When regions—plural— are engaged in a given function, we often use the term *network*. An example you will encounter in this book is the so-called default mode network (DMN): a set of mostly midline structures running from the front toward the back of the brain that tend to activate together under specific conditions, for example, states of relaxation and self-reflection, imaginative thinking, and engagement with stories (Chapter 6).

Neuroscientists have so far identified more than a half-dozen so-called large-scale, "intrinsic" cognitive networks defined in relation to differing tasks and/or states. So the training of neuroscientists and clinicians is shifting away from learning just the names—often from Latin sources—of various brain areas, and fiber tracts connecting them, to a system of nodes, connection pathways, and boundaries. The Latin names are still there in the background, but brain anatomy has definitely "gone cognitive" and is now based on larger, interconnected networks.

For a sense of how a cognitive systems framework is developing, here is a list of some of the key networks recognized from recent studies in functional imaging and connectomics (the science of defining the typical interconnections—the "wiring diagram"—of human brains and how those connections are adjusted during behavior),

- **Somato-motor network**—sensorimotor integration
- **Default mode network**—memory, imagination, self-reflection
- **Salience network**—emotion, some forms of pain

- Executive control network—decision making
- Dorsal attention network—localizing objects, events

These networks are not isolated from each other, but overlap structurally and interact functionally. The language describing large-scale brain networks frequently includes the terms "hub" and "spoke" that have technical meaning but also align with ordinary language. Just as airline networks commonly have a few central airports, hubs with many flights passing through (London, Frankfurt, New York, San Francisco, Sydney, Tokyo, etc.), there are connections extending to many smaller airports by way of spokes. When the central core of human cerebral cortex is mapped for large-scale organization, it actually does resemble airline route maps (e.g., see Hagmann et al. 2008, Bullmore 2016).

The functions associated with the brain's cognitive networks are not typically exclusive, so for example it would be incorrect to say emotion is regulated by only one of these and memory occurs only in another. Of vital relevance to the arguments in this book, language (aural and written) is a function that does not simply fall within one network. It engages regions of the dorsal attention network, but importantly engages others too, such as the default mode and salience networks. While the list of networks is evolving, it is fair to say that this new cognitive network view is a major contribution of neuroscience toward defining a physical basis for mind (e.g., Barrett and Satpute 2013, Medaglia et al. 2015, Peterson and Sporns 2015). Specific networks will be referred to ahead in the context of particular behaviors.

Brains as Cellular Networks

One of the most fundamental principles of neuroscience is that the basic structural/functional unit of nervous systems, and thus of their networks, is the individual nerve cell, or *neuron*. This is in keeping with the general biological principle that cells are the basic functional units of life.

We show neurons and DNA at the center of Figure 3.1, but will not have space to review genetics. Briefly though, each of our cells, whether neurons, skin cells, muscle fibers, etc., contains essentially every gene from our own genome, but only specific subsets of those genes are expressed (i.e., activated to make proteins) in cells of each type. So there are distinctive gene expression profiles for neurons compared to other cell types, and "the [human] brain expresses a greater number of genes than any other organ in the body. An additional indicator of brain complexity: it is composed of many distinct populations of neurons expressing different groups of genes" (Kandel et al. 2013: 41). Although genetics is a limited code, it can be compared to a language, which has a limited alphabet, limited grammar, but an almost unlimited number of sentences that can be made from these finite building blocks. Our life experiences, we now recognize, have a powerful effect on which genes are expressed (why arrows point from behavior to the brain and DNA in Figure 3.1).

Neurons, like other cells, are membrane-bound entities containing protoplasm (water, dissolved salts, and many larger molecules) and are surrounded by a salt water medium (extracellular fluid). The large molecules within protoplasm are diverse, but proteins should be mentioned because they are important for cell structure and also serve as enzymes—molecules that facilitate and control chemical reactions within cells. The thickest region of most neurons is still microscopic. The *cell body* (or *soma*), containing the nucleus and most of the DNA, typically ranges from a few *microns* to about 20 microns in diameter (1 micron = 1 millionth of a meter). So most are several times smaller than the diameter of a fine human hair.

Most nerve cells have even finer finger-like projections or processes extending from the soma and some of these ramify widely like the branches of a tree. These projections allow each neuron to be in "contact" with many other cells—at *synapses*—receiving bioelectric or chemical signals from perhaps thousands of others. Neurons perform operations on those signals, and then generate their own signals to be passed on to yet other neurons (or muscle cells or glands). One of the most characteristic signals used by neurons is the nerve impulse—which is crucial because it can transmit information at high fidelity over long distances.

It's common to distinguish between neuronal processes that receive inputs (*dendrites*) from those that transmit impulses (*axons*). Many neurons have a very large number of receptive dendrites, but typically only one axon. However, while this is the simplified textbook model (Figure 3.4 below), there are actually many alternatives. Neurons in different animals, and in different parts of human brains, vary widely in geometry. To the specialist, these geometries indicate different sorts of bioelectric signaling and information processing. The scope of these differences is beyond the present discussion but it is important to note that this structural dimension adds *a level of variability and complexity to neural information processing*. This phrase will recur in the paragraphs below, because there are many facets of the architecture of neurons themselves, and the circuits they form, that create a staggering array of possibilities for computational sophistication and dynamic adjustment.

The neuron in Figure 3.4 is intentionally generic, but in reality there are three very broad organizational classes. Neurons connecting directly to muscle via a special type of synapse are called *motor neurons*, while others that connect with sensory surfaces are called *sensory neurons*. In addition, neurons having direct synapses only with other neurons are called *interneurons*. Most of the nerve cells in our brain are interneurons and each typically connects to thousands of other neurons.

Neurobiologists have spent considerable effort studying neuronal membranes because they are integral to signaling and thus neural

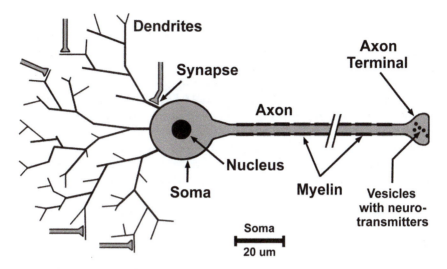

FIGURE 3.4 *Schematic drawing of neuron structure. In the dendritic tree, only the axon terminals of five other neurons are shown, but there might be thousands. This is not completely drawn to scale, but the scale bar shows that the soma is about 20 microns in diameter. Soma size varies, but the typical range of diameters in humans is about 5–20 microns. For reference, the diameter of human hairs range from about 50 microns to 100 microns.*

communication. For our purposes, it is sufficient to point out that membranes are incredibly thin, rich in lipids (fat), and contain specialized proteins that can control brief fluxes of ions (electrically charged components of dissolved salts) between the inside and outside of a cell. Individual body cells of *any sort* can do this; it allows them to sense aspects of their immediate environment and respond metabolically. However, in neurons the machinery of signaling is enhanced so that some membrane proteins exert exquisite control of ion flow across membranes; this results in digital pulses of bioelectricity that flow along membranes endowing neurons with the ability to encode and transmit information (for details see Aidley 1998, O'Shea 2006, Kandel et al. 2013).

The digital impulses of electricity referred to above are brief (perhaps 1/1,000 of a second in duration) and neurobiologists refer to them informally as *spikes*. They occur in other cells related to behavior too. In gland cells, spikes trigger hormone release. Some muscle fibers (cells) also use them, and their rapid spread along the fiber's length engages their contractile machinery in a simultaneous pulse of tension, or a "twitch."

Neuronal spikes sweep along an axonal membrane and can travel over long distances without losing signal strength. So what triggers spikes? First, sensory neurons respond directly to the energy of a stimulus (mechanical, photic, chemical) with physical changes in sensory receptors, changes that lead to small and very local ionic currents in the sensory zone of a neuron.

Those ionic currents may directly trigger a spike in the tip of a dendrite (such as a free nerve ending in the skin); or, if there is a discrete sensory cell (such as a taste bud), these currents can cause the sensory cell to synaptically activate an adjacent neuron to respond.

Think about a sensory neuron innervating the sole of your foot. A few impulses generated by a tickle there would have to travel from the cell's peripheral branch in the skin, all the way up its thin axon within a leg nerve to the spinal cord, where it activates other neurons. So, some neurons have to send signals over anatomically vast distances—not millimeters or centimeters, but on the order of a meter. One factor that helps convey impulses faithfully, with "high fidelity," over long distances is the reliability of biophysical mechanisms in neuronal membranes. On a macro-scale, in civil electrical distribution networks, this requires boosting stations to keep the electricity flowing properly.

Neurons in Context

In a human brain, there are at least as many non-neural cells—mostly *glial cells*—as neurons, and in some regions, more. At one time, glia were described as structural supporting elements and little more, but this is now known to be a seriously inadequate characterization. We've long been aware they form so-called myelin sheaths that wrap around some axons. Myelin is commonly found on axons involved in long-distance communication, acting as an electrical insulator, adding security, and speed, to spike conduction. It works as an insulator because it is highly enriched in lipids. This is one reason that brains have a high fat content. As you may know, loss of myelin insulation on nerves is associated with impulse conduction failures that cause the symptoms of multiple sclerosis.

More recently we know glia also regulate the chemical environment in which neurons are bathed. For example, sometimes they synthesize precursors needed by neurons to make chemical neurotransmitters (see below) and some have membrane channels that actively pump excess neurotransmitter away from synapses. This means they manage the availability of transmitters and the tempo of their action. Finally, some glial cells are actually mobile representatives of the immune system—patrolling the CNS for infections and abnormalities. We are still learning about the many important roles of glia and they certainly *add yet another level of variability and complexity to neuronal circuit function.*

Synaptic Diversity and Emergence

Every year reveals new and unexpected processes carried out by neurons (and glia). When neurons are interconnected they act as "circuits," allowing essential information about the state of our body and our environment to be shared widely in the CNS. Of course, it is a massive task to explain how one gets from a neuron response to systemwide functions like perception and memory—a scaling up that involves what neuroscientists call *emergence*.

For instance, if you were to focus only on one sensory cell in the ear, you could never predict that brain neurons respond to the subtle complexities of a human voice, nor could you predict we would be neurologically capable of analyzing syntax and meaning "on the fly" as words on a page are evaluated and understood within half a second or less; recently dubbed a word's "500 milliseconds of fame" (Wolf [2007] 2008). Reading and voice-recognition are behaviors we take for granted. But they show us that the work of humble neurons, in the aggregate, creates foundational aspects of our cognitive experience. For this reason, we will look, briefly, at the power of these remarkable cells.

Brains are, in part, communications networks. So, how are electrical impulses in one neuron relayed to another? When we look at synapses, the tips of neuronal processes are seen closely apposed to one another, yet there is a distinct gap, or "cleft," between their membranes. At chemical synapses, the electrical signals (usually spikes) passing along an axon mobilize the secretion of chemical *neurotransmitters* that diffuse through the narrow cleft between the *presynaptic* neuron and a *postsynaptic* neuron (or muscle cell, or gland cell). Diffusion over this short distance is rapid, and the binding of neurotransmitters to the membrane of the postsynaptic cell initiates a fresh round of bioelectrical signals. Although named by Sherrington after being inferred by him and others, we did not have definitive proof of their existence until the advent of electron microscopy in the 1950s allowed us to see the cleft and some associated structures. Later, synaptic physiology was analyzed in detail with improved electrophysiological recording methods and, more recently, with molecular biology.

There are a few places where synapses between neurons are not chemical, but rather electrical. Here, ionic currents that support spikes in one cell flow directly to the next, and the pattern of impulses in the *presynaptic* neuron is essentially repeated in the *postsynaptic*. Electrical synapses are often treated as minor players and ignored in summaries. But they show up in circuits where speed or security of cell-cell transmission is really important. For example, a sudden loud noise will typically cause you to jump slightly— and almost immediately—because of swift electrical conduction within the acoustic nerve to brainstem/spinal pathway: the initiation of prototypical escape behavior.

Chemical synapses have been targets of intense study because they are sites at which inputs to a neuron from various sources can be *integrated*, that is, combined or altered in meaningful ways. As bioelectric signals go, spikes are digital: they occur or not and are of a stereotyped size. However, there are local electrical events triggered by transmitter binding, called *synaptic potentials*, that are distinct from spikes. They are not digital, vary in size, and interact to add to, or subtract from, each other, before influencing the postsynaptic cell to "fire" spikes. If the aggregate synaptic electrical response is small (it can be measured as current or voltage) it may have no effect on spike generation. However, if the total synaptic response from all active inputs to a neuron reaches a threshold level, then a spike, or a train of them, will be triggered. Chemical synapses are the most common type in our brain.

There is a vast array of chemical transmitters, and each region of the brain tends toward synapses associated with a certain type. You might say synapses come in many different chemical "flavors." Also, some transmitters increase the probability of postsynaptic impulses being fired and are called *excitatory*, while others that decrease the probability of firing are classified as *inhibitory*.

One way to imagine the range of possibilities here is to realize that not only do transmitters come in a wide variety of distinct "flavors," but some individual neurons use more than one type of transmitter, so their synaptic flavors are complex, chemical "cocktails." In addition to all of this neurotransmitter diversity, there are some generalized signaling chemicals (*neuromodulators*) that in liquid or gaseous form bathe whole areas of brain. Their presence can have powerful effects on how transmitters do their job. The neuromodulators are a bit like the elements on a dinner table that augment flavors: well-chosen spices, or sauces. With all these possibilities, no two "neuronal banquets" are likely to be exactly the same.

Neuronal "Negotiations" and "Decisions"

In a recent review of how the ideas of neuroscience contribute to our understanding of human communication and cognition, Gregory Hickok (2014) explained how, after the waning of behaviorism—which essentially denied that cognition mattered—linguistics, and especially the work of Noam Chomsky, provided an alternative: "The [linguistic] system is productive: given a basic set of categories and computational rules, it can generate an endless stream of combinations ... Today we call this approach to cognition the *computational theory of mind*, the idea that the brain is an information processing device that performs computations" (114).

Whatever else our brain may do, it certainly does carry out computations. We can say that a single neuron is both a constituent of living tissue *and* a computational node. It is important to define what we mean by this. Essentially, the computational dimension can be understood by considering synapses through the analogy of negotiation and decision-making. This allows appreciation of the brain's levels of intricacy and the potential for "emergent properties" within neural networks.

The interaction of thousands of opposing synaptic influences impinging on any neuron, and the fact that they come with positive (excitatory) and negative (inhibitory) signs is the first factor that generates complexity

(Figure 3.5A). If synaptic electrical responses overlap spatially and/or temporally, they can combine: excitatory synapses can enhance each other's effects and inhibitory synapses can diminish the impact of excitatory synapses: a compromise. You can think of the initial, local synaptic electrical response as a preliminary "decision" by the neuron. That preliminary level of excitation may lead to an impulse (a definitive decision) *only* if it reaches a required threshold at the "spike initiation zone" where a cell's axon begins (cross-hatched area in the figure).

Because the local synaptic electrical events that interact in dendrites vary continuously in size, they are *analog* processes. Such interactions occur over periods of a few milliseconds to perhaps a few seconds. The ongoing interplay of excitatory and inhibitory synaptic responses thus leads to different degrees of compromise between the inputs over time—and subtle modulation of the spike sequences that the responding neuron generates. The rate of impulse generation is thus sculpted across time by the waxing and waning of excitatory and inhibitory inputs.

As if this isn't complex enough, an additional level of structural intricacy derives from synaptic arrangements, as shown in Figure 3.5B. Sometimes synapses are formed on other synapses in a nested relationship. This allows

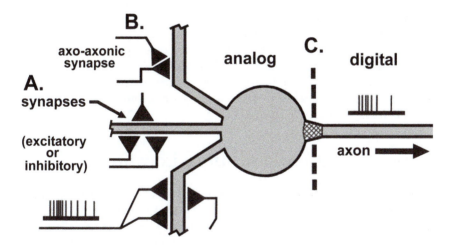

FIGURE 3.5 *Most neurons act as multi-compartment devices. This shows a schematic neuron with a number of synapses. The triangles are exaggerated denotations of synaptic terminals. Spike firing is shown for the postsynaptic neuron and one presynaptic cell (dark baseline with vertical lines indicating a hypothetical sequence of impulses). Sources of complexity: A: There are many thousands of synapses and they may be excitatory or inhibitory. B: There can be 'synapses-on-synapses' for fine modulation of chemical transmission. C: The many signals arriving interact with each other electrically in the postsynaptic cell's dendrites: their analog synaptic currents are integrated at the initiation zone (cross-hatched) setting up patterns of digital spikes in the cell's axon.*

the synapse to be modulated: either increasing its effect on the postsynaptic cell (*facilitation*) or decreasing that effect (*depression*).

Now, consider the outcome of such synaptic "negotiations" on neuronal signaling. As indicated, spikes come in one size only; therefore, they are *digital* events and it is their temporal pattern—frequency—that conveys information. For example, in sensory cells, weak stimuli are invariably conveyed by a low spike frequency, but stronger stimuli produce higher spike frequencies.

The perhaps surprising upshot is that each neuron is a "hybrid" processing device: displaying analog processing for the negotiation of synaptic influences, but a digital process for expressing final decisions (Figure 3.5C). The ability to work in both modes facilitates further permutations and nuancing: flexible negotiation during the analog phase and unambiguous decision-making after translation to the digital mode.

Figure 3.5 summarizes the three ways in which the possibilities for "computation" by neurons can become extremely intricate. Ultimately, neurons can interact in ways that amount to addition, subtraction, multiplication, and division. The net effect is: a given pattern of impulses in one neuron can lead to a varied pattern of impulses in another, depending upon the nature of pre- and postsynaptic modulatory influences.

Neurons, then, can perform not only the basic arithmetic functions, but also the operations of calculus—some networks even seem to perform vector math—and this is not an exhaustive list of the enormous possibilities. In short, circuits of neurons are capable of a wide range of intricately nuanced negotiations—occurring on the scale of thousandths of a second to a few seconds or, sometimes, minutes. To discover the many elegant mathematical and engineering approaches to brain and synaptic physiology, see Arbib (2003) and Dayan and Abbott (2005).

Location, Location, Location

Location is fundamentally important in neuronal circuits. Consideration of inhibition really makes this clear: inhibitory synapses will have different levels of influence on spike generation simply by virtue of their location (closer to the spike initiation zone is usually more potent). Also, a distinction needs to be made between presynaptic and postsynaptic inhibition. Presynaptic inhibition, via nested synapses, can selectively dampen one channel of inputs compared to others and potentially change the balance of information a neuron gets to work with in a very finely graded manner. Alternatively, a postsynaptic inhibitory connection can, when located near the spike initiation zone for example, act like a "master switch" and attenuate the influence of most or all of the inputs arriving throughout the dendritic processes. Needless to say, there is a vast range of modulation or dampening between these two extremes.

Despite the reality of analog and digital processing in neurons, and the many levels of neuronal and circuit complexity, the prevalent model for the brain remains a digital computer. That model is sorely deficient. Our brain is not fixed hardware like a computer's processing unit; rather, it is interactive, adaptable "wetware," and it is amazing that such small bags with salts, proteins, etc. dissolved in water can perform combinatorial dynamics that lead to immense cognitive power.

Yet, neural design at both cellular and systems levels combines to produce highly varied and subtly modulated forms of information processing. To summarize some of the factors behind this: (1) neurons are intrinsically complex processors using interactive analog and digital capabilities; (2) their geometry is highly differentiated and that correlates with a range of signal processing styles; (3) neurons have numerous, and not yet fully understood, functional relationships with glial cells; (4) synapses may be electrical or chemical in nature; (5) chemical synapses come in many distinct chemical varieties, and are subject to a range of biochemical modulators; and (6) the locations and spatial relationships of excitatory and inhibitory synapses allow for a wide array of integrative modes or "algorithms" that can be enacted in neuronal circuits. Humility is required of us: such a list does not come close to all of the possibilities.

As a result, when thousands or millions of neurons are formed into circuits, modulated by glial cells and bathed in varying chemical cocktails, the functioning of a whole circuit may—and typically does—display properties not found in any individual cell: hence, we find *emergent properties*. Some theorists even believe consciousness itself as an emergent property. This is fine as far as it goes: to describe any property of the nervous system as emergent may be a start, but it does not even come close to explaining the conscious state in all its protean depth and subjectivity (for examples related to, and found within literature, see Chapters 5–10).

Plasticity and Learning

The multifactorial dance of excitation and inhibition at synapses not only is a crucial element in the scientific narrative of how cognitive functions emerge, but also facilitates the modification of functions over time. There's no doubt that learning, remembering, and forgetting involve changes at synapses, based on biochemical alterations in the "negotiations" within and between cells. Not surprisingly, many potent drugs that affect the brain do so by altering one or more components of the synaptic molecular machinery.

The term *plasticity* is widely, and often loosely, applied to the phenomena of learning and compensation, even though these processes are quite distinct. Normally, it describes the responsiveness of neural systems to changing conditions, to include developmental processes in young organisms, collateral sprouting from axons after nerve damage, or the adjustment of cognitive structures after a physiological "insult" such as a stroke. The term is something of a catch-all: maturational changes, adjustments for pathology, but also learning, tend to fall under its scope. However, they are quite different in their details. Science does not always avoid linguistic ambiguity.

A rich area explored in studies of plasticity is how map-like brain structures are organized or adjusted within sensory and motor systems. Early work by neuroscientist Michael Merzenich and his colleagues showed that in response to peripheral lesions, or stroke, the primary somatosensory areas of primates adjust their representation of intact sensory surfaces, at the expense of lost or "denervated" territory. Naturally, this phenomenon is clinically important for recovery. However, of more relevance to quotidian brain function, it was observed that when perfectly intact monkeys were required to repeatedly use certain digits to touch textured surfaces for rewards, the area related to those digits expanded in the somatosensory cortex. "Rewiring" in this case—abandoning some synapses, building new ones—is influenced by chemical neuromodulators of the cortex, and is also influenced by the timing of sensory signals in neurons. For example, when two fingers are used independently in behavior, they maintain separate areas within somatosensory cortex; but if used synchronously in monkeys or in people, their representations in the cortex can become indistinct (Jenkins et al. 1990, Wang et al. 1995).

Such use-dependent plasticity is well-known in the case of musicians. Over time, violinist's brains show a spread of the cortical area for the hand used to finger notes. Similar adjustments have also been reported in association with tool use. The intriguing conclusion is this: sensory, motor, and cognitive structures at the cortical level were once thought of as fixed, especially in adults. However, it is clear that something as basic as our first-order sensory structures reflect usage, and this leads to the conviction that our brain reflects a sum of two forces: both unfolding developmental programs *and* our day-to-day experiences. While sensorimotor and cognitive structures occur at more or less predictable sites in our brain, they are nonetheless *idiosyncratically organized* and reflect our unique personal history (e.g., Elbert et al. 1995, Schaefer et al. 2004, Schwenkreis et al. 2007). The changes in prefrontal cortex (Chapter 2) that are associated with the ability to perceive the intentionality of others are a pertinent example.

Most learning is expressed at a physical level by alterations in synapse physiology and structure. In this regard, one of the compelling stories to come out of the latter half of the twentieth century was the discovery of how alterations in the use of synapses translate into enduring changes at molecular and cellular levels. A few paragraphs of summary cannot do justice to some of the elegant mechanisms involved, but two contributions are particularly relevant to cognition.

First, Canadian psychologist Donald Hebb, in his book *The Organization of Behavior* (1949), re-cast classical processes of learning by locating them at specific synapses. His predictive model, or compact narrative if you will, was epitomized in the punchy maxim, neurons that "fire together will wire together." As a result, such associative processes happen at what are now called Hebbian synapses, found at multiple sites in mammalian brains.

Second, the processes of modulating the strength of synapses and their *specificity* (how just a few relevant synapses are modulated on a neuron that may have many thousands) have also been studied. Eric Kandel has been very important in this field, receiving a Nobel Prize in 2000. He and his colleagues studied learning at defined synapses and their results have wide implications, not just in explaining how a sea slug or a mouse can engage not only in simple non-associative learning, but also in more elaborate "associative" learning.

His book *In Search of Memory* (2006) is a classic. It summarizes his contributions, and those of many others who deserve recognition, through a series of well-written narratives about various phases of discovery. Not only is the research explained clearly, but the personal experiences of the researchers are intertwined to make the account doubly compelling. Kandel has been an eloquent voice in both his technical articles and his books, reminding us that erudition, a value we often associate with humanists, is no less important in science. Also recommended is a fine volume that Kandel produced (2012) on the synergies between art and science traceable to turn–of-the-century Vienna. Here again, he has personal connections to the story that are deftly woven into the overall narrative.

Sociality

Mind, in a practical, moment-by-moment sense, may be, above all, an operational model we use to predict, follow, and respond to the pattern of another individual's behavior. Without doubt, the lineage of primate evolution made us a highly social species in which the complex interactions within and between groups have a profound effect on individual survival and chances for leaving offspring. Anthropology and brain science concur that successful social living requires special skills that can be considered components of "fitness." Managing such skills in ever-larger groups has been hypothesized as a factor driving the increased brain size within our lineage (e.g., Dunbar and Schultz 2007, Silk 2007).

This is an area of rich literary exploration, and has given rise to the specialized field of evolutionary psychology, which appears to offer some answers. A balance is required, though: it is ill-advised to try and explain all of human psychology on the basis of selective forces we faced in the Pleistocene, yet it is equally undesirable to assume that evolutionary forces lack any explanatory potential. Evolutionary hypotheses are useful for answering "why questions" about traits (their origins and persistence), even if they are of much less help for answering "how questions" (their mechanisms). Furthermore, there are abundant indicators that our evolution did not magically stop 100,000 to 200,000 years ago.

A pressing challenge for neuroscience and cognitive studies is to explain what components of group behavior may be unique to us, how brains influence each other under social contexts, and how we developed the cognitive capacity to understand and even internally model the intentions of others.

Neuroscience has been providing new clues at an accelerating rate about how our brains respond to social cues (see especially Chapters 2 and 7). Many of these highlight the significance of oral language, gestures, facial expressions, and an understanding of intentionality. In addition, parallels with close primate relatives turn out to be really useful in showing us what is unique about human culture (Kovacs et al. 2010, Stephens et al. 2010, Stanley and Adolphs 2013, Hari et al. 2015).

The Flexible Brain and the Impact of Culture

A pressing question in social neuroscience is how culture may in turn influence human brains and minds. Plasticity, learning, and memory are neural processes that must play important roles in how culture (and its products like narrative and literature) impact our mind. They undoubtedly provided means by which humans absorbed stories and oral histories before writing and formal education emerged; thus they allowed the earliest narratives to become durable learning tools, cultural norms, and sometimes works of art.

A recent finding gives these general ideas convincing and detailed corroboration. As we will see in Chapter 6, it has been known for a while that although speech is, at its core, an inherited capacity, reading is a learned skill. Through the application of a wide array of traditional and newer techniques such as functional brain imaging, several research groups, that of Stanislas Dehaene and others, have pinpointed a location in the brain where visual circuits are modified by experience, so that object-recognition mechanisms eventually become devoted to the comprehension of letters on a page. This is an outstanding example of how recent neuroscience is uncovering the substrates of our "literary brain."

Such work has practical importance too, as it helps us understand the mechanisms that may undermine normal reading development. Most importantly, it also shows how a social practice, the teaching of reading, causes observable physical changes in children's brains that enable them to share in the narrative heritage of their culture. The arrival of culture must have radically changed the evolutionary environment. And in considering this, an unavoidable conclusion emerges: just as the nature versus nurture debate turned out to be a false dichotomy, so too is that of biology versus culture (Byars et al. 2010, Courtiol et al. 2012, Courtiol et al. 2013). This point will be expanded in the chapters ahead.

Notes

1   Two recent examples: Will Self, *The Guardian*, May 2, 2014; Tom McCarthy, *The Guardian*, March 7, 2015.
2   In addition to selection, gene frequencies may also change because of random "drift." For a rich discussion of relevant evolution-related processes, see Daniel Dennett (1995) or Zimmer and Emlen (2013).
3   This accessible book on brains and behavior summarizes the Reith Lecture of 1976 and contains a useful historical sketch of early ideas on mind. Podcasts of these lectures are available at http://www.bbc.co.uk/podcasts/series/rla76/all.
4   See Robert Graves (1973) and Crivellato and Ribatti (2007) on the heart, and other organs, as a presumed site of mental activity.

4

Deep Substrates of Narrative Imagination

Early in the twentieth century, a common simile for the human mind was the high technology of that era—a telephone exchange. Charles Sherrington (1906) had been one of many to apply this to the integrative capacities of the brain. However in *The Principles*, I. A. Richards' opinion was clear: in order to understand the complexities of imaginative states, "The chief misconception which prevents progress here is the switchboard view of the mind" (1924: 234). By the time Sherrington summarized his views on the brain in 1940, he had aligned with Richards' approach instead describing the transition from sleep to wakefulness in one of the most enduring metaphorical images of the brain:

> The great topmost sheet of the mass, that where hardly a light had twinkled or moved, becomes now a sparkling field of rhythmic flashing points with trains of travelling sparks hurrying hither and tither. The brain is waking and with it the mind is returning. It is as if the Milky Way entered upon some cosmic dance. Swiftly the head-mass becomes an *enchanted loom* where millions of flashing shuttles weave a dissolving pattern always a meaningful pattern though never an abiding one; a shifting harmony of subpatterns. (Sherrington [1940] 2009: 225)

There is elegance in this view. It comes close to describing how we currently understand our brain, constantly buzzing with activity. Aptly enough, it also happens to resonate strongly with Elaine Scarry's subsequent suggestion that "radiant ignition" is a powerful mechanism for the imaginative evocation of motion.

Modern functional imaging, ultimately traceable back to Sherrington's research, has updated the picture he presented. Our brains hum along at high throttle across levels of consciousness, under anesthesia, and even through some parts of sleep (it is not as "dark" as Sherrington's description implies). This led a pioneer of functional imaging in the recent era, Marcus Raichle (2011), to speak in almost parallel terms of "the restless brain." In this regard, Sherrington's imagery of "flashing points" and "travelling sparks" nicely captures the high energy electrical signaling in the brain: indeed neural activity patterns are not "abiding" (even functional connections between regions wax and wane in sync with behavior—something only appreciated recently with newer scanning techniques).

Sherrington's cosmic comparison suggests a universe within. The concept of a loom (high technology of earlier centuries) links human goals to their realization in rhythmic patterns forming and resolving into subpatterns, as fabric both practical and beautiful is assembled. This description is close to how we now understand cerebral dynamics. The loom metaphor appeals too because it suggests the cognitive art of weaving stories.

Metaphors Still Matter

For all its age, Sherrington's analogy makes current metaphors of the brain as computer seem anemic. Furthermore, his metaphor has also influenced contemporary discussions of the cross-currents between literature and science. A. S. Byatt acknowledged her debt to Sherrington in 2004, "My own writing and thinking have been much influenced by Sir Charles Sherrington's metaphors for mind and brain." In the same interview, she acknowledges the impact of current neuroscientist Antonio Damasio who will recur in our discussion because his name has often been invoked by contemporary novelists, "I am naturally sympathetic to Damasio, who speaks to my response to Sherrington's metaphors" (Byatt, *The Guardian*, February 14, 2004).

The metaphors we choose speak volumes about how we struggle to understand challenging concepts. Near the end of the twentieth century, a book titled *The 3-Pound Universe* (Hooper and Teresi 1986) provided a summary of the surge in neurobiological research that was getting underway, using a metaphor that has been repeated and recast over time.

Within the past few years, our brain was described as a "3-pound control center" in a book by neuroscientist David Eagleman (2011) looking at how much of our behavior is actually conscious (not so much). In counterpoint to Eagleman, author Brian Appleyard, uses the three-pound trope to convey a skeptical stance: in *The Brain Is Wider than the Sky* (2011), he refers to the brain as "1.3 kilograms [about three pounds] of mainly fat and water" (65). The thrust of Appleyard's book is to question whether science can give us meaningful descriptions of mental phenomena by studying brains using physical techniques. His reductive label, three pounds of fat and water, used repeatedly in the text, reinforces the impression that attempting to explain our mental life by studying the brain is futile. But what, really, is the description implying? It's true that human brains generally weigh about three pounds—and have a high percentage of fat (lipids) and water. Our brain is about 60 percent fat by weight, other organs much less; the body, overall, is mostly water. But this metaphor is akin to characterizing a jumbo jet as merely 160,000 kilograms of metal and plastic, or a Puccini opera as two hours of noise.

The structural complexities of brains at a cellular level are staggering (Chapter 3) and render characterizations based on gross chemistry inadequate, even spurious. They hide the intricacy of brains, where multiple *levels of organization* coexist. Given neurobiological complexity, the time has come to consider the substrates for language and narrative at levels from single neurons to large networks. The evidence in chapters ahead shows that operations that emerge from multiple brain levels, together with our social/ cultural environment, are *both* necessary to comprehend our predisposition for stories. Below we begin considering the neural components that will ultimately provide insights into our story-hungry, social and emotional minds.

A Multiplicity of Visual Systems

We start with the sensory system that has been the most studied, and which is essential to reading and writing: vision. Classic studies have demonstrated repeatedly that, counter-intuitively, the "visual brain" is not a single entity: we actually have several functional visual systems differentiated according to behavioral tasks. In Figure 4.1, arrows indicate the path of visual information traveling from the retina at the back of the eye along the optic nerve.

Some signals go directly to the midbrain *optic tectum* (called *superior colliculus* in mammals). A wealth of experimental evidence demonstrates that the tectum contains a topographic map of the retina, and is involved in orienting behavior: it directs eye movements with respect to the location of an object of interest. Often this is toward a positive stimulus such as food, but it may be away from a threatening stimulus. This is spatial localization, and so the tectum can be called a "where" pathway. Yet spatial information about stimuli is also processed at the cortical level (see below): so why are there two pathways for it in the CNS?

One answer, in people at least, has to do with levels of cognition. The operations of the midbrain are generally not immediately accessible to consciousness, and so it's more like an automatic channel of control, whereas much of what happens at the cortical level does enter awareness. Having an automatic visuo-motor loop ensures rapid reactions to many stimuli. The cortical level is the province of more considered responses. So, we really have "two visual systems": one is a midbrain system for detecting objects and reorienting our gaze, the other a cortical system that goes beyond localization to analyze fine features of scenes/objects (Schneider 1969, Ingle 1973).[1] In what follows, we emphasize the cortical level.

Looking at Figure 4.1 again, notice that in addition to retinal signals going to the tectum, some optic information also passes, in parallel, through the thalamus and then to visual cortex. The thalamus is buried beneath the cortex and one of its functions is as a gateway for the flow of signals into and out of the massive cortex above. Also notice: from the primary visual cortex at the back (occipital lobe), there are two arrows projecting forward. That's because there is another bifurcation within cortical visual networks: a lower (ventral stream) "what" pathway and an upper (dorsal stream) "where" or "how" pathway.

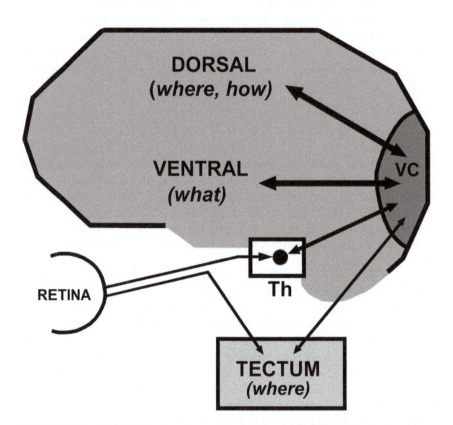

FIGURE 4.1 *Multiple brain pathways for vision. This schematic drawing shows that the path from the eye to the midbrain tectum is direct; a parallel path from eye to visual cortex (VC) is indirect, relaying through the thalamus (Th). Visual cortex is the origin of at least two distinct pathways going forward in the cortex: one for identifying objects (the* what *pathway), and another for locating stimuli and directing motor responses (*where, how *pathway). The visual cortex and tectum (another* where *system) are interconnected, note most arrows are drawn to reflect that two-way traffic is typical in these pathways.*

The dorsal stream is called a "where" pathway because, like the tectum, it can encode locations of objects, and this information is used for guiding our conscious (and some unconscious) directional movements: providing upper level guidance of gaze, reaching movements of the arms, and some control of grasping. Damage in the dorsal pathway can lead to optic ataxia, or impairment in visually guided reaching, and so this pathway is in general associated with conscious and skilled movements (e.g., Farah 2003), as distinct from movements generated on "autopilot" such as by the tectum.

To turn to a related and equally important question: how do we identify objects? After all, this knowledge determines whether we turn toward, away from, or ignore objects; it is a vital activity dependent upon the ventral stream in the cortex. This "what" pathway allows visual signals to percolate through areas beyond the occipital lobe and into the inferior temporal lobe where there is an amazing set of modules specialized for visual identification. These modules are where visual features of faces and bodies are "mapped," where various sorts of visual environments we know are represented, and where different types of "objects" are systematically categorized by type: animate, inanimate, tools, and, crucially, text.

These areas in the ventral (what) pathway might be called conceptual maps, or cognitive structures. We know from neuropsychological data that damage in this area can lead to selective visual agnosia (impairment in recognizing things seen) of various types, for example, damage to a face-selective region can produce face blindness, damage in a tool selective area can lead to confusion in naming tools (e.g., Reddy and Kanwisher 2006, Ungerleider and Bell 2011, Konkle and Caramazza 2013). In relation to reading, the ventral stream is especially important to the process of form-recognition we use to identify letters and words. This will be dealt with in the next chapter.

Neurons as "Gnostic Units"

At the cellular level, the functional organization of visual systems has real implications for understanding the underpinnings of narrative imagination, and this requires going beyond simple studies of sensation to answer more complex questions about visual cognition. Early studies tried to understand vision first by recording the response of retinal and brain neurons to spots of light. Then in a famous paper entitled "What the Frog's Eye Tells the Frog's Brain," Jerome Lettvin's group at MIT asked how responses were created to natural (and moving) objects. They described several classes of cells projecting from the eye into the optic tectum of frogs. Some of these were sensitive to visual borders and others discharged impulses selectively for erratically moving dark spots. This latter type became known prosaically as "bug detectors" (Jerome Lettvin et al. 1959, and see Grobstein et al. 1983, Roche King and Comer 1996, Ewert 1997, for correlations with behavior).

In a similar way, classic studies of mammalian cortical visual systems by David Hubel and Torsten Wiesel found neurons visually responsive to light or dark bars—often with selectivity for their orientation. These are often called "edge detectors" and they've been characterized in primate brains. Any given visual neuron has a field of view—its "receptive field"—depending on what part of the retina it is "connected" to. And they display selectivity. For example, the edge detectors are strongly activated (fire many spikes) for a dark or light edge within its receptive field, often even being selective for the orientation of the bar (vertical, horizontal, oblique, etc.). Still others require moving bars or edges to become active (Hubel and Wiesel 1968, 1977). These neurons seem to be early building blocks for encoding the perceptual experiences we have of objects and even people. You might then expect some cortical neurons to be selectively activated by convergent input from subsets of edge detectors so that we can recognize shapes. Of course some neuroscientists followed this logic too.

Neurophysiologist Jerzy Konorski was very influenced by the work of Hubel and Wiesel and their description of basic edge detectors ("simple cells") up to progressively more sophisticated detectors of higher order visual features ("complex cells"). Perhaps, he reasoned, these would be integrated to produce neurons selective for real-world objects. In his 1967 book *Integrative Activity of the Brain*, he predicted that neurons would be discovered with selective responses for faces, hands, and perhaps emotional expressions. The name he gave such cells was "gnostic" neurons (from the Greek for knowledge). Konorski also predicted that specific cortical areas for gnostic neurons would be found. This was a bold prediction and sparked many debates among neurobiologists in the 1960s and 1970s.

Such debates led Jerry Lettvin from bug detectors to the proposed existence of a "grandmother cell," the idea that our recognition of a person might depend upon one, or a very few, unique cells attuned to their visual characteristics. It was more of a "straw man" than a serious theory, but it is an example of how important stories can be to thinking through, and teaching, issues in brain science. Charles Gross (2002) has described how the idea came from a story told by Lettvin in a class on perception:

When discussing the problem of how neurons can represent individual objects, [Lettvin] told a (tall) tale of how the neurosurgeon A. Akakhievitch [his second cousin] had located a group of brain cells that "responded uniquely only to a mother ... whether animate or stuffed, seen from before or behind, upside down or on a diagonal or offered by caricature, photograph or abstraction." At this point, Lettvin introduced the mother-obsessed character from Philip Roth's (1969) novel *Portnoy's Complaint* and Akakhievitch ablated all of the mother cells in Portnoy's brain. As a result, Portnoy completely lost the concept of his mother. So, "Akakhievitch then went on to the study of grandmother cells."

(512)

Thus the term "grandmother" cell came to refer to a neuron, or a few neurons, that responded only to specific, complex stimuli and thus represented that percept or concept.

The idea of gnostic units, or grandmother cells, is one extreme on a possible spectrum of neural coding: suggesting perhaps "one cell for one object." Contrast this with large *ensemble coding* in which we only have neurons responding to certain lower level features such as lines or edges of certain orientations—perhaps they could signal the angle of a nose, the shape of a chin, etc. However, this would necessitate the discharge of many such basic "feature detectors" to encode the face of your grandmother, not just one cell. This alternative became the accepted norm in mammalian neurophysiology.

Charlie Gross has said that Konorski's speculations and Lettvin's fable both influenced him to embark on a quest for, if not a grandmother cell, at least neurons with higher-order properties. Sure enough, Gross and his colleagues went on to find face-selective neurons in 1972. It is hard now to appreciate just how radical their findings were.

Cells with such dramatic specificity, almost matching our lived experience of faces, had never been seen before (Figure 4.2, from one of a series of

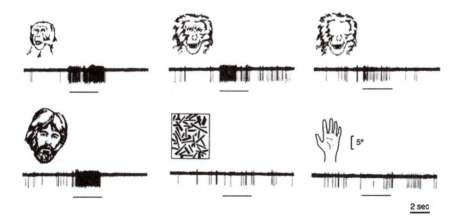

FIGURE 4.2 *Neurons in temporal cortex of macaque monkeys responding selectively to faces. Each panel within the figure shows a stimulus at top and approximately 12 seconds of neural activity from one isolated neuron below. Stimuli were projected on a screen in front of the animal near the center of gaze. The neural impulses here are the vertical lines directed downward from baseline. Note the difference between weak responses (bottom center and right) and the strong responses, with so many impulses that the lines run together. From: Bruce et al. (1981). Visual properties of neurons in a polysensory area of superior temporal sulcus of the Macaque.* Journal of Neurophysiology, 46:369–84. *Figure 7. We include six or the nine panels in their original figure. Used with permission.*

papers by Gross and colleagues). Most neuroscientists probably recall a handful of papers across a whole career that really inspired a sense of awe about what the brain can do. This was one of them. After all, the question is at least as old as Aristotle: how do we create a reliable sense of objects, and their associated concepts, in our heads? This isn't straightforward. After all, we recognize people we know under different lighting conditions, when seen from varied angles, when dressed differently, etc. Even today, programming facial-recognition software to do the same task is so complex that the most advanced technology we have is forced to combine three-dimensional recognition, skin-textural analysis, and algorithms for extracting "landmarks" from facial features, to be compared against a database.

Ultimately, what Gross found in the temporal lobe was not a single cell that would encode "grandmother" or "uncle Joe," or even a large ensemble of these, but rather an example of a process called *sparse coding*: activity in a modest collection of cells of the type he found could encode an individual. He concluded that a small group of neurons is capable, through combined firing, of evoking recognition of people you know—the spike discharges providing the internally "whispered" neural signature of their identity. Sparse coding has now been reported in several brain locations in monkeys, and in humans (e.g., Gross 2008, Tsao et al. 2006, Kanwisher 2006).

As will become apparent, cells with the complexity of face cells, and sparse coding, are essential background for cognitive functions like memory, reading (Chapters 5 and 6), and even imagination.

Sensation and "Small Spatial Stories"

As noted, Mark Turner (1996) believes the human mind is "essentially literary ... [and that] Narrative imagining—story—is the fundamental instrument of thought." He goes on to explain how fundamental it is by comparing it to vision: "We almost never notice the activity of vision or think of vision as an activity, but if we do, we must recognize that the activity of vision is constant and more important than anything we happen to see ... only a neurobiologist is likely to notice the constant mechanisms of vision."

Turner extends his analogy: "Story *as a mental activity* [our italics] is similarly constant yet unnoticed, and more important than any particular story." He goes on to identify a fundamental unit or "cognitive primitive" underlying it:

The basic stories we know best are small stories of events in space: The wind blows clouds through the sky, a child throws a rock, a mother pours milk into a glass, a whale swims through the water. ... these small spatial stories may seem hopelessly boring. We are highly interested in our coherent personal experiences, which are the product of thinking with small spatial stories, but we are not interested in the small spatial stories themselves. (12–13)

There are several interesting overlaps between Turner's approach and basic information from neuroscience and cognitive linguistics. To begin with "small spatial stories," in Turner's opinion:

1   Are executed without conscious attention,
2   Are what a human being has instead of chaotic experience,
3   Are so necessary that they are pre-cultural,
4   Yet they are the core of culture. (1996: 14–15)

To be precise, he is referring to the basic processes of sensation and perception on one hand, and to our linguistic representation of these processes on the other. We can relate this to visual perception. Recall Figure 4.1. Vision happens once signals from the eyes reach the tectal or cortical levels. Both levels contain fundamentally spatial cognitive structures: that is, contain topographic maps of the retina and hence "spatiotopic" maps of the visual world (because the world is optically projected on the retinal surface by our lens). Both levels maintain the spatial coherence implicit in the retina: this is one way that our visual processing is not "chaotic." Using Turner's example, if "a child throws a rock," its trajectory would quickly activate different retinal regions and corresponding regions in cortical (or tectal) "maps" of visual space as a result of its flight. Since we all possess these pathways, their operations—as far as they go—are "pre-cultural." So, in what sense can they be the "core of culture?" The answer would seem to be language. Here is one way that might work.

The linguist Ray Jackendoff points out even simple sentences have what he calls a "spatial structure." For example, one can think of spatial structure as an image of the scene described by the sentence. This is an intuitively appealing notion for neuroscience and for exploring the link between perception and language. When considering how conceptual patterns emerge from semantic processing, Jackendoff says: "These patterns are the basic machinery that permits complex thought to be formulated and basic entailments to be derived in any [linguistic] domain. Among these domains, the spatial domain exhibits a certain degree of primacy due to its evolutionary priority and its strong linkage to perception" (2002: 358–9). This point is similar to Elaine Scarry's about the connections of language to imagining: in most cases, humans perceive the world within an inherently spatial framework. Thus a spatial framework exists as a primal level of meaning inherent within language.

Vision enables communication, emotional expression, reading, and the arts. We all use these mechanisms to notice a cloud move across our visual field, or track a thrown rock, or predict where a whale may surface next. It seems fair to suggest that these perceptual activities are part of the deep neural background for formulating and understanding stories.

Neural Systems Intimately Connected to Speech

The complexity of language means our brains must be capable of integrating information from visual pathways, auditory pathways, motor systems, and memory to engage in producing and understanding speech and other forms of language. While there is overlap and some convergence of these systems, research over two centuries has demonstrated that there is no single site in the brain where they converge and language takes root. Here we are considering language as communication by means of words with agreed-upon meanings, whether vocal or written. To be sure other factors—like posture, facial expressions, and gestures—are often part of our communicative repertoire, but of less interest at this point.

There was a time when it was common to say that language is controlled by the left cerebral hemisphere, which is "dominant" in this respect. Yet this is not a clear or precise characterization of the situation as will emerge from material in this chapter and the next.

What's actually clear is that the control of speech (the ability to articulate the sounds of the vocal domain of language) is highly developed on the left side of the brain. This is an old observation, and important as one of the earliest demonstrations of a connection between a brain structure and a cognitive skill. The French neurologist Pierre Paul Broca (1824–80) studied a patient, Mr. LeBorgne, who had suffered a stroke resulting in severe *aphasia*—restricted or lost speech production, but with some ability to understand the speech of others—called an *expressive aphasia*. On autopsy, a lesion was found on the lower lateral aspect of the left frontal lobe (Figure 4.3A). Eventually, it was established in a small group with similar symptoms that the language-altering lesions were all in essentially the same part of the frontal lobe, usually on the left side (Broca [1865] 1986, Rorden and Karnath 2004). Lesions in the equivalent area on the right side usually do not produce aphasia.

Remarkably, because Leborgne's brain was preserved, it was possible to scan it (via magnetic resonance imaging—MRI) after more than 150 years and verify lesion location with more accuracy. This modern study concluded that the lesion was in the cortical location originally described, but that

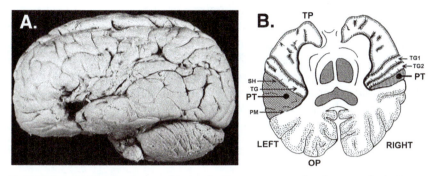

FIGURE 4.3 *Observations on human brain asymmetry related to speech. A: Picture of the brain of Mr. Leborgne (looking at left hemisphere, front of the brain is to the left) as observed at autopsy by P. Broca in 1861 (see Broca 1865). B: Drawing of a roughly horizontal slice through a human brain. It runs from the rear (occipital pole, OP) on a slight downward angle through the temporal lobe toward the front (temporal pole, TP). The posterior margin (PM) of the planum temporale (PT, shaded) slopes backward more sharply on the left. The anterior margin of the planum formed by the sulcus of Heschl (SH) slopes forward more sharply on the left. There is a single transverse gyrus of Heschl (the site of primary auditory cortex) on the left, but two on the right (normal variability). Left: from Signoret et al. (1984) Rediscovery of Leborne's brain: Anatomical description with CT scan.* Brain and Language *22:303–19. Figure 2. Right: from Geshwind and Levitsky (1968) Human brain: Left-right asymmetries in temporal speech region.* Science *161:186–7. Part of Figure 1. Both are used with permission.*

damage below the cortical surface also was substantial. This doesn't mean Broca got it wrong, but rather that the system he identified is more extensive than originally assumed (Dronkers et al. 2007).

Broca's finding of lateralization for speech was interpreted to mean that in the normal operation of the brain production of words and phrases typically resides in the left cerebral hemisphere, with the right hemisphere only rarely in control. This aspect of the work has stood the test of time and it was the beginning of the idea that the left hemisphere is "vocal" and "dominant" for speech, while the right came to be called "silent."

Then in 1875 Karl Wernicke reported another region dealing with language function, now known as *Wernicke's area,* in the temporal lobe near its junction with the parietal lobe that includes the *planum temporale* (Figure 4.3B). Lesions to this area produced a distinct deficit, not of speech production but of speech comprehension: a so-called receptive aphasia. In the left hemisphere, the *planum* (an important sound analysis area adjacent to primary auditory cortex) is distinctly larger than on the right—an anatomical asymmetry presumably related to importance of the left *planum temporale* in hearing and speech. So both speech "centers" known by the start of the twentieth century were lateralized predominantly to the left side of the brain.

Given the classic observations of Broca and Wernicke, a model for vocal language focused on strongly lateralized speech control has been in textbooks for many years (Figure 4.4A). Heard speech was assumed to be decoded for understanding in Wernicke's area and then formulated into articulate words and phrases in Broca's area. Because there are known fiber pathways that connect these two areas (primarily the *arcuate fasciculus*), it suggested a simple circuit with two nodes: one primarily sensory, the other motor. Inevitably, the reality is more complicated than a reflex-like loop of this kind.

Speech Assessment in Neurosurgery

A more realistic view of speech control in the left hemisphere now exists.[2] Accurate maps of the territory controlling speech are indispensable, not least for the operating theater. Neurosurgery is often done while the patient is awake so that behavior (especially speech) can be assessed during its course. The allure of a physician reaching into the head, touching the brain, and perhaps influencing an aware mind has not been lost on literature. In 1621, it was mentioned in Thomas Burton's *Anatomy of Melancholy*. In the recent era we have seen, a flourishing of narratives, novels, and non-fiction, about or by writers like Lawrence Shainberg 1979, 1988, Ian McEwan 2005a, Karl Ove Knausgaard 2015, and Henry Marsh 2015, 2018.

Neurosurgeons must know in a practical way where key functions related to language are likely to "reside" if they are to be protected when operating inside the skull. A recent review written primarily for neurosurgeons provides a clear update on the classic model for brain control of speech functions (Chang, Raygor, and Berger 2015). Figure 4.4 compares the traditional model mentioned above and this recently updated version (panel B).

Some features of the traditional model are noteworthy. First, following Wernicke's original identification, his area is described as rendering a "sensory word image," which is somehow translated to a "motor word image." Activations in Broca's region create the discharges of both upper level (motor cortical) and lower level (brainstem) motor neurons required to move laryngeal and tongue muscles for speech. As the diagram shows, diffuse connections from these two hubs with surrounding cortex were assumed to link the two language zones to conceptual processing, memory, emotional associations, etc.

Much of today's understanding came from studies by neurosurgeon Wilder Penfield, who had worked for a time with Charles Sherrington before he moved to Canada and directed the Montreal Neurological Institute. There he conducted a series of remarkable studies while operating on the brains of conscious patients (under local anesthesia). With localized electrical stimulation, he could selectively interfere with tasks such as naming objects or matching nouns to verbs. Such temporary disruptions identified so-called eloquent areas to be carefully avoided when, for instance, removing a tumor.[3]

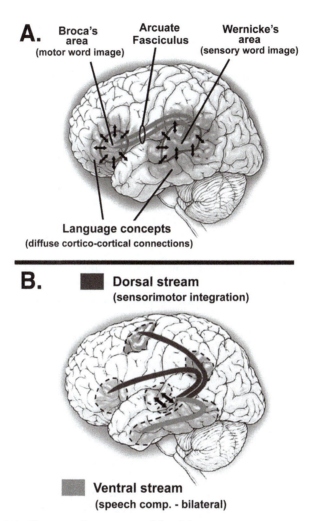

FIGURE 4.4 *Classic and recent model of language organization in the left hemisphere.* **Top:** *Classic model: Broca's area is in the inferior frontal lobe; Wernicke's is in the posterior, superior temporal lobe. They are connected by a fiber pathway, the arcuate fasciculus. Language concepts are assumed to be represented in areas surrounding each language area.* **Bottom:** *More recent dual stream model of language. Shaded regions interconnected by short double arrows show initial cortical processing in zones around primary auditory cortex. The ventral stream (*what *pathway, light gray) flows into the temporal lobe (lowest shaded region), and is involved in the comprehension and representation of lexical concepts. The dorsal stream (*how *pathway, black) passes through the parieto-temporal boundary, then one branch connects to pre-motor cortex (top-most shaded region) and another to the inferior frontal gyrus (most rostral shaded region), thought to bring phonological information to articulatory control centers. Note all arrows/pathway markers are bidirectional. From: Chang et al. (2015). Contemporary model of language organization: An overview for neurosurgeons. J.* Neurosurgery *122:250–61. Figures 1 and 2. Enhanced for legibility from a color version, used with permission of the author.*

One thing that Penfield and colleagues noted was that "in a given individual, the location of essential language sites can be extremely variable and nearly impossible to predict preoperatively" (Chang et al. 2015: 253). The individual variation in language localization is insightful: the cortex has a typical organization, but as with any biological structure there is a range of possibilities. As neurosurgeon George Ojemann pointed out, variability in the sites at which functions such as speech, lexical memory, and reading are situated suggests that each person's cortex is, at a fine level, idiosyncratically organized for linguistic and cognitive tasks (Ojemann 1991). This view is fully consistent with the information in Chapter 3 on the demonstrated plasticity of sensory and motor maps as a result of life experience.

It was not until the past ten to fifteen years that a more complex picture of the circuitry for speech and language emerged, coming not only from clinical observations, but from new brain imaging techniques and careful behavioral testing.

You can see in Figure 4.4 panel B, that additional pathways are now recognized. Remember that visual function was more completely understood by the realization that there are distinct *ventral* and *dorsal streams* for, respectively, identifying and acting upon visual objects. A similar dichotomy was found in the pathways for speech and language (e.g., Hickok and Poeppel 2007).

After sound information arrives in primary auditory cortex, speech components are quickly processed in superior temporal regions—Wernicke's area. Then the information enters a ventral stream (*what* pathway) projecting throughout the temporal lobe, where speech comprehension begins: the sound components of words are identified in a process termed *phonological* analysis and there must be a lexicon of some sort to give us word meanings (the rostral pole of the temporal lobe has been consistently linked to *semantic* analysis). However, a dorsal stream (*how pathway*) runs through the parietal lobe and then forward to the motor and premotor cortices. This is where our speech movements are controlled.

Inevitably, as Chang et al. (2015) indicate, the details of the pathways as we now know them are more complex than shown in Figure 4.4B. For example, in auditory cortex itself, where sound processing starts, observations show that it contains distinct sub-regions specialized for sound localization *and* sound identity, including a zone selectively sensitive to sounds with the qualities of the human voice (Moerel et al. 2012, Ahveninen et al. 2013, Jasmin et al. 2019). More surprising was that new analytical techniques applied to functional MRI data showed another zone in auditory cortex tuned with high selectivity for music (Norman-Haignere et al. 2015). This raises an old debate about the cognitive relationship of speech and music. Their representation early in the handling of sound information by the brain—within mostly distinct, adjacent zones—suggests that music must have been of significance to humankind for quite some time. Darwin would smile, as he always assumed this was the case.

The Distinction between Speech and Language Is Revealing

If speech shows a strong bias toward left-side control, does this mean that language does too? Is the right hemisphere after all "minor" and "silent"? Most recent findings are uncovering a more competent right hemisphere than we were led to believe twenty years ago. Here is one way to see it: while the output path to speaking is very left-lateralized, and the initial inputs of sound and visual information (heard and read speech) are quite left-lateralized, the great *cognitive middle ground* of language, broadly construed, is more bilateral.

Although lateralization remains important, it is by no means the entire story. One neurophysiologist working in the area has announced, "The age of Broca's and Wernicke's areas and the era of left- hemisphere imperialism are over" (Poeppel 2014). Both hemispheres (LH and RH) are involved in directing the flow of information from the temporal lobe to other cortical, and subcortical, locations and this involves a half-dozen fiber pathways, including some only recently identified (Dick et al. 2014).

However, much information still comes from careful analysis of patients with lesions to one side or the other, and as we will explore below, from "split-brain" patients. In his detailed analysis of such data, shedding light on how our two hemispheres "work," Iain McGilchrist (2009) has said that instead of a left-right dichotomy, "the hemispheres are evolutionary twins: they display a remarkable degree of overlap or redundancy of function, and run in parallel rather than in series" (10). But how can we then characterize the main contributions of each?

Well, McGilchrist very usefully summarized the findings from many studies indicating that the language capacity of the LH is highly literal, whereas the RH "processes the non-literal aspects of language" (49). So the LH is good at logical analysis and inference, and these abilities are disrupted by LH lesions. Conversely, those with RH lesions have difficulty understanding metaphors and don't "get" jokes. Because humor so often relies on non-literal and contextual aspects of language they are lost on the LH. In addition, appreciating the emotional prosody (rhythmic patterns) of speech (Kreitewolf et al. 2014) and voice qualities that may reveal speaker identity (Belin et al. 2000, Perrachione et al. 2011, Kuhl 2015) is largely RH dependent—both vital factors in our social and emotional lives.

Cognitive neuroscientist Michael Gazzaniga, a major contributor to "split-brain" studies, pointed to hemispheric differences in more mechanistic terms: "RH language has a different organizational structure compared with LH language" (2000: 1313). For instance, the RH is not capable in the area of phonological analysis, nor is it able to manage with most grammar, but it does have a lexicon. It can usually understand individual words and act accordingly, even if it can't directly say what it knows except by way of the motor system of the LH, although occasionally people are found who can speak from the RH, and those with LH lesions begin to do so during recovery.

Finally, when using "natural" or more "ecological" stimuli—that is, narratives—it's been found that the RH is more strongly activated than the LH for tasks like generating proper endings for sentences, deriving themes, detecting story inconsistencies, and determining the sequence of narrative events (see summary in Beeman 2005). These abilities are crucial for what most would consider a rich and full language capacity. As we will see in the next two chapters, the use of whole narratives as stimuli in recent studies, along with new ways to extract data from brain scans, is providing insights that would be impossible to appreciate from analysis of words or phrases in isolation. This is one way new roads into narrative comprehension are opening up to us.

Lateralization and Narrative

A provocative suggestion about narrative function in our brain has come from Michael Gazzaniga and colleagues' observations on "split-brain" patients, where the cerebral hemispheres have been isolated by cutting the fibers of the *corpus callosum* that normally interconnect them. At one time, this was done for relief in cases of intractable epilepsy. After this radical surgery, it was possible for the first time to interrogate each side of the brain and probe its distinct cognitive-linguistic propensities.

What at first seemed startling was the general lack of behavioral consequences to this massive alteration of brain connectivity: no notable shifts in overall behavior were initially observed! Subsequently, extensive tests verified that severing the callosum "blocks the interhemispheric transfer of perceptual, sensory, motor, gnostic and other forms of information in a dramatic way" (Gazzaniga 2005: 653). This line of experiments revealed that the hemispheres could operate independently, with unique forms of information processing, attention mechanisms, and even seemingly distinct personalities.

In Gazzaniga's experiments, a surprising aspect of what we might call *narrative drive* was unexpectedly revealed. In early studies, he flashed images to either the left or right sides of the visual field or to both simultaneously, and asked simply: "What did you see?" Patients could describe only what was projected to their right visual field (seen by the left cerebral cortex]) This was no surprise because (a) the pathway from eye to cortex is crossed such that each side of visual space is represented in the opposite cerebral lobe, (b) speech production most often displays LH lateralization, and (c) the callosum is the main pathway by which the left side of visual space, seen by the RH, has access to speech production mechanisms in the LH.

However, when the patients were given tests where their responses consisted of pointing, not speaking *and* then asked why they pointed to particular objects, they could "speak" about choices made by the left hand—that is, with the RH that was not able to articulate for itself what it had seen. The pattern of results led to a breakthrough:

> We gave each hand a multiple-choice option. The left hand was free to choose one of four pictures that best matched the left visual field stimulus. The right hand was free to choose one of another four options to match stimuli presented in the right visual field. With the task set up in this fashion, we could change our question. Instead of asking what the patient had seen, we would let each hand respond and then ask, "Why did you do that?" (Gazzaniga 2013: 13)

The outcome with a patient identified as PS on this "simultaneous concept" test illustrates the dramatic results (Figure 4.5).

A scene with snow was shown to his left visual field ("non speaking" RH) and that of a chicken claw was shown to his right visual field ("vocal" LH). Each cerebral lobe has a "crossed" mode of control for limbs, directing movement only on the opposite side. When asked from an array of pictures, to pick something relevant to what he had just seen, the patient pointed to two pictures: each a conceptual fit with what that isolated cerebral lobe had viewed. The left hand (controlled by RH) chose a snow shovel and the right hand (controlled by LH) chose a chicken. Asked to explain these choices, a snow shovel and a chicken's claw, the left (verbal) hemisphere (not having seen snow) said "Oh, that's simple. The chicken claw goes with the chicken, and you need a shovel to clean out the chicken shed" (Gazzaniga 2000: 1316).

So the left hemisphere improvised—constructed a small story of sorts to explain this unusual left hand choice (snow shovel), given the only visual input available to it (chicken). In short, the left hemisphere came up with a confabulated narrative justification. Gazzaniga (and colleague Joe LeDoux) concluded, "There appears to be a special module in the left hemisphere [they dubbed 'the interpreter'] that makes up an explanation for why all the [brain] modules do what they do" (Gazzaniga 2013: 13).

In many experiments, they observed the "interpreter" freely confabulate to make retrospective sense of any unusual sensory input *and* discordant emotional responses. He summarized the unexpected implications:

> There it was. It took 25 years to ask the right question, and in doing so, perhaps the most important finding from all of split-brain research was revealed. One of our seemingly infinite modules *generates the storyline* as to why we do the things we do, feel the things we feel, and see the patterns in our behavior that contribute to *our theory about ourselves*. Once you see it at work in this simple experiment, you see it everywhere. The responses from discrete modules pour out of all of us, and evolution invented a module to make it all seem like it pours forth from a "self." (Gazzaniga 2013: 14, our italics)

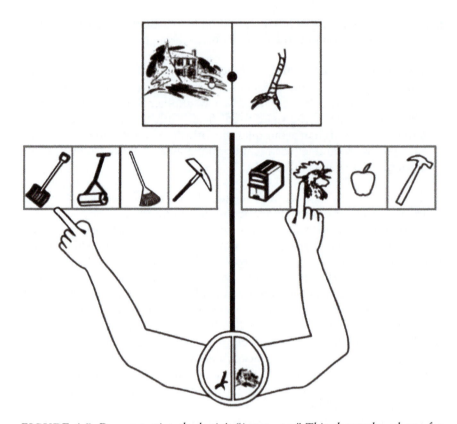

FIGURE 4.5 *Demonstrating the brain's "interpreter." This shows the scheme for the simultaneous concept test: arrangement used to test patient PS, who had all connections between the cerebral hemispheres cut. Both sensory and motor pathways are crossed at the cortical level: information from the two eyes is sorted out so that, with the eyes fixated straight ahead, the right hemisphere only has input from the left sensory field (and controls the left hand). The left hemisphere only has input from the right sensory field (and controls the right hand). From: Gazzaniga (2000) Cerebral specialization and interhemispheric communication. Brain 123:1293–326. Part of Figure 19. Used with permission.*

These clinical tests are consistent with the idea Mark Turner called "narrative imagining," part of his idea that narrative is fundamental to the human cognitive kit. He sees the production of stories as a core mental activity of humans: we construct them incessantly. As William Hirstein (2006) has noted, the brain routinely and subconsciously confabulates when required, creating "brain fictions," which show up conspicuously in a number of clinical syndromes.

It is important to note these experiments tell us something about what the left hemisphere can do in terms of speaking. They do *not* provide evidence that the right hemisphere lacks a role in narrative construction. As the work on confabulation makes clear, what you conclude about a part of the brain depends upon how you "ask it questions." Equipped now with at least a rough "map" of where some basic functions may be found in our brain, the next two chapters explore a more integrated view of how our evolved capacity for speech and communication can be molded by cultural forces to enable understanding of the natural and social world through narrative.

Notes

1   The information leaving our eyes goes to more targets in the brain than we discuss here. So there are really more than "two" visual systems. These other "systems" depend upon other CNS locations and they control things like circadian timing, neuroendocrine function, and some types of eye movements.

2   A nice summary can be found in McGilchrist (2009) "In the west at present, about 89 percent of people are broadly right-handed and the vast majority have speech and the semantic language centers in the left hemisphere" (12). If you consider that many left-handed people also have speech in the left hemisphere, then this pattern of lateralization is present in 90–95 percent of the population. The remainder either have right side lateralization, or bilateral control.

3   In *Do Androids Dream of Electric Sheep*—the science fiction novel by Philip K. Dick that inspired the film *Blade Runner*—Penfield receives a "nod." In this post-apocalyptic world emotions are important because they distinguish the remaining humans from androids. Ironically, an induced emotional lift is available in homes with a device called a Penfield mood organ.

PART III

The Journey from Words to Narratives

In the next two chapters, we consider some theories about how words "came to be" during the deep history of our species, and how the origin of stories maps onto that same history. We also look at the neurobiology of language, and give an account of how science expanded its view from a highly reductionist focus on words in isolation, to a recent explosion in the ability to study words under natural conditions, as part of oral and written narratives.

5

Compelled by Words

In an interview, the author Tessa Hadley was asked what moved her most in a work of literature. Her reply: "Is it just too obvious to say, the words? Because those come first, rather than the themes or the stories or the analysis. When you're standing in a bookshop opening up a novel or book of short stories by someone you've never read before, there's a period of rapid testing, where you enter into the sentences" (*New York Times*, January 3, 2016).

That view on the primacy of words is paralleled by the thoughts of critic Stanley Fish, who described his reaction to a radio interview with novelist Colm Toibin in which Toibin dodged attempts to characterize a recent book of his as a response to the death of his mother and brother. Fish was, at first, annoyed by Toibin's refusal to connect with the emotional pitch of the questions, but then reconsidered: "What I had first regarded as a suspect evasiveness was in fact a determination to be faithful to the practice he was dedicated to and a refusal to claim for that practice effects that could not or should not be its objective." Fish noted that Toibin's questioner seemed to be searching for an external justification for the writing, such as providing solace. The truth was much simpler. Toibin insisted that assuaging grief, or offering any other kind of consolation, is not what writing is about. The act of writing makes use of grief as it might make use of anything. As Fish put it, "Toibin was saying, I write because making things out of words is what I feel compelled to do" (*New York Times*, February 11, 2007).

In the previous section we approached language quasi-historically so the emphasis was naturally on speech. We also considered some larger issues in our mental handling of speech and language in terms of processing at the level of the cerebral cortex, noting that some functions—like speech production—are strongly but not absolutely lateralized to the left hemisphere. We also outlined how language functions show up at predictable places in our brains, but with idiosyncratic variations—so the cerebral topography of language belongs uniquely to each of us. Both hemispheres must be involved in any complete understanding of language, that is, its handling both literally and with the inevitable figurative, humorous, and contextual features.

In this chapter and the next, we travel deeper into lexical territory before stepping back to view the broader terrain where words occur naturally: real narratives that people listen to and read. Finding a way for neuroscience to engage with narratives is vital and it has been aided by technology and innovative methods to chart the working brain in real time as stories are created and received.

It seems natural to use words as a starting point for a consideration of language and reading, brain and culture. It is also possible, of course, to dig into the technicalities around syllables, phonemes, and other "prelexical" components. However, a proper analysis at that level would take us far from our primary interest: the behavior of those who listen, write, or read, so we will touch on those underlying details only briefly.

The priority in this chapter is to explore how writing and reading differ from speech. Reading and speech are distinct in origin and history; they are each handled in characteristic (yet overlapping) ways in our brains. We also will describe how cognitive systems for reading are not developed entirely de-novo within a child's brain but are created, with a major prompt from culture, within a brain that has a pre-existing competence for oral language.

Human Family History

Figure 5.1 shows a general scheme for thinking about humankind in "deep history." By this phrase we include the earliest era with historical-archaeological records and the evolutionary era where we were still in the process of becoming fully modern humans (anatomically, but especially, behaviorally).

On the order of 6–10 million years ago, our immediate lineage (the hominins) became distinct from the primate populations that were precursors to modern gorillas and chimps. The details of this split are only vaguely known. However, the hominin lineage was essentially a "natural experiment" in bipedal locomotion, producing ambling, striding, and running types (like us). The importance of bipedalism, it is hypothesized, was primarily that the hands were available for selection toward tool use and gestural communication. This evolutionary "experiment" with body and behavior was associated with a concomitant trend toward increasing brain size.

From the most recent half-million years of the fossil record, there are two main hominin types we know in considerable detail: our own, *Homo sapiens*, and the Neanderthals, *Homo neanderthalensis*. As you can see (Figure 5.1A), our ancestors overlapped with the Neanderthals, who eventually became extinct. Several other hominin types with whom we co-existed (not shown in the figure) have been discovered within the past fifteen years or so; these include *Homo naledi*, *Homo denisova*, and *Homo floresiensis*.

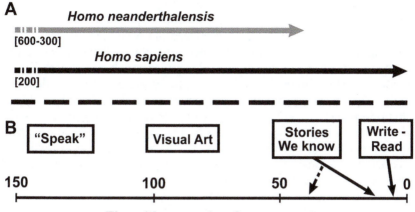

FIGURE 5.1 *Schematic of recent human evolution and major behavioral innovations. A: Shows the most recent several hundred thousand years: the approximate times when* H. neanderthalensis *and* H. sapiens *became distinct lineages are given in brackets to the left. There were other hominin species as described in the text. B: The time of origin for speech as a communicative system is almost certainly more than 100 thousand years ago. The origin of visual art is now usually placed between 40,000 and 100,000 years ago. The origin of writing/reading is known from archaeology and history to be about 5,000 years ago. The origin of stories is much more speculative. Oral tales may have origins from about 5,000 to perhaps 15,000 years ago as described in the text (solid arrow). However, to the extent that some early visual art suggests cognitive sophistication, and may have had a narrative function, storytelling may be closer to 50,000 years old (dotted arrow); whatever its age, oral storytelling clearly preceded writing/reading by many millennia.*

Bipedal locomotion resulted in not only mobility on the savannahs of Africa and greater use of the hands, but also group cooperation in hunting, and the provision of more varied and high-quality food. The new diet most likely enabled the expansion of brain size within the hominins, beginning in Africa. The earliest hominins, Australopithecines (not shown in Figure 5.1), had brains about 450–500cc (cubic centimeters), only slightly larger than current chimps, and they did not leave Africa. Subsequently, a hominin type referred to as *Homo erectus* is represented in the fossil record: they were larger in stature, clearly hunted game, and had brains ranging from 800–1,000cc. Their "burst" in brain growth happened somewhere between 2 and 1.5 million years ago, and they survived up until perhaps 100,000 years ago. There is fossil evidence of *H. erectus* across parts of Eurasia, following migration out of Africa.

The largest hominins with the biggest brains are the recent homo species, shown in Figure 5.1A. *Homo neanderthalensis* and *Homo sapiens* arose about 600,000 and 200,000 years ago, respectively. Brain volumes of Neanderthals and *H. sapiens* ranged from about 1,400–1,600cc.

By about 40,000 years ago, we displaced the Neanderthals throughout their range and they went extinct (Chapter 2 touches on a narrative view of this piece of deep history). To add a wrinkle, genetic analyses reveal that there was some interbreeding between early *Homo sapiens* and the Neanderthals: indeed 1–4 percent of most Eurasian genes are traceable to them. So, a degree of hybridization occurred, at least outside Africa. Thus, not only was the early human family tree crowded but its branches intimately intertwined. The Neanderthal genes Eurasians acquired have been traced to influences on skin, immune function, and neurological or psychiatric characteristics of those possessing them (Simoniti et al. 2016).

The Historical Border between Writing and Visual Arts

Figure 5.1B displays some behavioral milestones relevant to discussions of language and literature from a bio-cultural perspective. To describe this "sequence," we start at the present and work backward: moving from linguistic issues with clear historical evidence, and cautiously entering prehistory where the evidence is incomplete, and less compelling. Notice first that writing and reading are to the far right: they are known to have origins in human cultures about 5,000–6,000 years ago. However, speaking— verbal language—is to the far left, arising as far back as 100,000–150,000 years ago. That is a conservative guess, but one informed by multiple sources of information, some of which we will touch upon. Importantly, the temporal separation of these two human capacities is correlated with a foundational premise for which there is substantial evidence: speaking evolved biologically and is rooted in our genetics. Writing and reading are recent, and importantly, are cultural inventions.

These crucial facts must cause us to ask about the conditions for the origin of writing and reading. Humankind began adopting a less nomadic lifestyle and engaging in agriculture approximately 10,000 years ago. Dating back to about 4,000–3,000 years BCE, there are specimens of clay "tokens" bearing symbolic markings found in association with human settlements in the Near East. These were initially used to tabulate sales of land, and commodities such as bread, beer, and cattle. Maryanne Wolf, in her book *Proust and the Squid* ([2007] 2008), speculates that these tokens were a form of proto-writing that may have engaged some of our ancestors in a form of proto-reading.

We know that these tokens were often kept in clay "envelopes" with markings on the outside as to their contents, and the envelopes ultimately became clay tablets with rows and columns for the accounting of commodities, debts, etc. (Schmandt-Besserat 1981). Careful analysis of a corpus of clay tablets from the Near East indicates that the lined tablets gradually shifted from a numerical and economic function to a lexical one, perhaps influencing the way ideas were expressed.

Art at this same time, roughly 3,500 BCE, shifted from clustered or juxtaposed motifs, for example, some geometric, some of animals, to linear arrays that could *tell a story*. So, with the advent of literacy, "pottery, seals, stone vases, steles, and wall paintings, changed from evocative to narrative," or in other words, art apparently followed writing's lead. This lexical-to-art influence is proposed to be part of a larger reciprocal interaction where writing influenced art and then the further development of art influenced writing (Schmandt-Besserat 2007: 59).

From this exciting work, we can now say that literate writing emerged by about 3,000 BCE in Mesopotamia (currently part of Iraq). At almost the same time the hieroglyphic writing system in Egypt began to emerge. Alphabetic systems such as those of the Greeks probably go back to about 1,000 BCE (Wolf 2008).

The Formative Ecology of Narrative

Many scholars of literary history believe that our written stories display evidence that they were originally developed as part of an oral tradition, and only later inscribed in clay or on papyrus. For example, it has been pointed out that early historical references indicate the stories attributed to Homer were present in early Greek civilization as memorized poems, probably sung. Triangulating from a few such references, it has been estimated that the Homeric epics go back to about 800 BCE—a time when alphabetic systems were barely gaining ground (see Boyd 2009: 215–18 for a concise description of timing, and controversies around authorship). So it is possible that the oral traditions of these epics were in place for several generations before being written down.

Were there earlier stories? There must have been: one possible window on this was opened by studies of myths and traditional tales. There are systematic collections of such stories across multiple language groups put together by anthropologists, linguists, and folklorists. Biology has recently contributed methods that some anthropologists have begun to use. Here is the logic for tracing the lineage of stories: when a tale shows up in similar formats across several language groups, a reasonable assumption is that the story existed originally in the language that was the "ancestor" of the various cultural/linguistic traditions sampled. This would assign a potential minimum age to the story. This is the same logic that uses changes in the DNA of genes to assign relationships, and times of splitting, between species. To be sure, the adoption of such "phylogenetic" methods to study story evolution is not "settled science." It involves various levels of statistical tests, and sometimes the recognition of "mythemes," or story components, that some believe may shift and evolve over time, as genes do.[1] While conclusions drawn from using these methods are still speculative, they nonetheless provide hypotheses that are intriguing and worth pursuing via other approaches.

For instance, "phylogenetic" models of this type can be checked against analyses of the ages of languages based on the frequency of common words across linguistic divides, and from genetic data on the interrelatedness of various populations and their migrations. Some examples suggest the potential value of these studies. One case comes from the attempt to trace some of the numerous stories associated with the Indo-European family of languages. In particular, the fable of "The Smith and the Devil" (believed by some to be distantly related to the tale of Faust) shows up—in a very general form—in many of these languages and is thus hypothesized to have originated in about 5,000–5,500 BCE. It would therefore be coincident with the Neolithic or very early Bronze Age, or about as old as the entire Indo-European family of languages (da Silva and Tehrani 2016, Pagel 2016).

In another example, d'Huy and colleagues have analyzed the "Polyphemus myth," which occurs as an episode in Homer's Odyssey. In their study, they have inferred it is connected to wide-ranging stories about a "master of animals," and trickery related to freeing captive animals. From this broader perspective, it might be related to stories told by Native North Americans, such as the "Trickster Crow." If true, the data on when these populations first left Asia and crossed the Bering Strait produce the astounding conclusion that the origins of this particular story are on the order of 15,000 BCE—quite different from the consensus of earlier scholars, and very long before there is evidence of written language (d'Huy 2016).

These emerging, and still controversial, data provide a reason for the approximate chronology for the origin of "Stories We Know" in Figure 5.1. The purpose of this diagram is not to assign, or argue for, a specific time at which stories emerged, but to make clear that there are reasons to think, just from analysis of such stories and language development, that narratives in some form might have been part of our "cognitive play" for 10,000 years before writing emerged. As we will describe below, findings of early visual art could push the estimate back even deeper in time.

Because of its likely relevance to cognitive development, we also show a tentative origin for visual art in Figure 5.1B, although its roots and significance have their own complex and rapidly changing narrative in paleoanthropology. For many decades, the origin of artifacts that our modern eyes would recognize as art has been given as 30,000–40,000 BCE. However, recent findings in Africa have challenged how we define the word "art," and are pushing its probable origin back toward 70,000–100,000 BCE (Henshilwood et al. 2011). While the beginning of visual art is fascinating in its own right, it also lends a perspective that some believe helps map the temporal origins for speech and narratives.

Much evidence comes from representational art: well-known cave paintings in Europe and Asia from around 30,000–40,000 years ago and even earlier artifacts indicating work with pigments, together with ornamental carvings from Africa that date to 70,000–100,000 years ago (e.g., Henshilwood et al. 2002, 2011, Aubert et al. 2018, Henshilwood et al. 2018).

But how would we recognize visual art that at least gives the intuitive feel of narrative cognition? One recent suggestion is to focus on prehistoric cave paintings that depict scenes. Maxime Aubert and colleagues (2019) have presented such an argument recently from analysis of scenes found in caves on the Indonesian island of Sulawesi, which may depict episodes of hunting, and were chemically dated to a minimum age of 44,000 years (perhaps twice as old as other scene paintings, although it's a tricky conclusion—many cave walls were in effect lithic palimpsests with figures altered or added over time). However, if some of those they have reported are indeed coherent scenes, the inclusion of figures with mixed animal and human form (*therianthropes*) adds an element that suggests the paintings represent sequential elements of imaginative stories. Art that we can recognize as such now exists from multiple sites around the world for the period of about 15,000–45,000 years ago, and the recent find from Indonesia is by no means the only one that is depictive in nature. Our guess is that these artworks should be taken as proxies for a measure of symbolic thought and—perhaps a bit bigger leap—narrative thought and language.

For all the discussion of our direct forebears, we should not ignore the Neanderthals. We overlapped with them and interbred; it's natural to wonder about their mental world. Certainly they lived socially, and there is good evidence of artistic activities, taken of late to suggest they exhibited some imaginative cognition (Appenzeller 2013, Rendu et al. 2014). Golding's novel *The Inheritors* derives much of its haunting power by articulating their perspective.

The Murkiest Historical Map: The Origins of Speech

Speech is the oldest of our unique behaviors displayed in Figure 5.1B. There are several reasons for the consensus opinion that oral language is an evolved capacity. The ability to speak is found in all human populations, and it displays "spontaneous emergence"—all children, reared under normal conditions, attempt to speak. Thus it is a *specieswide adaptation*: an inherited cognitive capacity that helps define what it means to be human. If it evolved, there must be a genetic basis for it, and indeed some genes for speech have been identified. The best-known is called *FOXP2*. All modern humans have a version of it, and individuals with mutant copies have deficits in articulation and grammar.[2] We know that Neanderthals carried the same form of *FOXP2* as we do. This gene is necessary for speech, but we do not know what genes if any are sufficient for it, so this does not tell us if Neanderthals spoke or indeed when we acquired that ability. We can look then to evidence from paleoanthropology.

Ideas on oral language origins from the past century include hypotheses about its derivation from gestures of the body and face, animal calls and/or emotional cries, grooming and gossip, or even—as Darwin implied—singing. Always ahead of the field, Darwin proposed that while speech is instinctive, reading and writing are not ([1871] 1977: 464), and he appreciated that there are natural connections between speech and emotion.

Words: Rules, Meanings, Levels of Understanding

Words are a natural starting point for speech and language study generally, so it is useful to pause and describe their cardinal properties. First they are associated with sound(s)—this is how we experience them in our earliest childhood. Next, it is clear that even a single word may have a propositional character. When an infant says "cup," it may name something, but it may equally be a question (cup?) that will, in time, be qualified with additional words. Finally, words are assembled as sentences by way of a grammar, a set of rules about proper word sounds, how they are effectively ordered in phrases, and their agreed upon meanings all of which we internalize. We will focus on words and phrases largely as they embody meaning: not least because recent neuroscience has made intriguing progress on understanding how our brain organizes our semantic lexicon, our inner library of word meanings.

Having drawn attention to the central importance of words, we also need to consider their power in sentences. With a finite number of words as building blocks, we can combine them to make a nearly infinite number of sentences. Furthermore, we often nest phrases inside others: "Once the sun sank toward the horizon, the children, who had been swimming, dried off and sat on the deck." This syntactic nesting, called *recursion*, helps explain the nearly infinite meanings that can be constructed from a finite repertoire of words. It is not only important to language as we know it, but it captures some of the complexity of human thought more generally (Hauser et al. 2002, Corballis 2007, Bolhuis et al. 2014).

In Chapter 3 (Figure 3.1), we drew attention to the fact that neuroscience treats the complexities of brain, body, and environment by working systematically at several levels: while neurons are always a key focus, there are separate methods and theories for working "downward" to sub-cellular components like molecules, or working "upward" toward networks, body physiology, and environment or culture. In a roughly parallel way, narrative study has a key focus on words, but sometimes must focus "downward" to sub-lexical components like phonemes and morphemes, or perhaps more, often focus "upward" to sentences, narratives, texts, and environment or culture—with differing methods and theories for each level. The point of noting this rough operational sense in which words are "like" neurons is to acknowledge that narrative studies are in their own right as multi-level and complex as those of neuroscience. For interesting consideration of the multiple levels related to language and narratives see Emmott ([1997] 1999) and Jackendoff (2012).

Word and Phrase Origins

Despite wide agreement that oral language is an evolved capacity, there is a lack of agreement on when it emerged and how it came to be. Opinions tend to diverge in two main ways: some believe it arose perhaps 200,000 years ago, or more, and assumed its current form gradually as it flourished in support of communication and social relationships. Others have argued it is much more recent, perhaps 50,000 years old and arose rapidly accounting for complex features of our grammar—like recursion.

If oral language is indeed quite old, then our primate relatives provide some clues about possible precursors. Words might have been based on simple calls—rather like the "wrrs" and "chutters" of vervet monkeys or the "pant hoots" of chimps (e.g., Seyfarth and Cheney 1992, Kojima et al. 2003). This would likely have been speech, we could call a *protolanguage,* with only rudimentary symbolism.

This sort of scenario, less concerned with language as thought and prioritizing language's communicative role, is one that anthropologist Richard Leakey—among others—prefers. The assumption of such models is that language came "online" gradually and probably included gestures and mimetic posturing. This approach allows us to see language in its full biological and psychological complexity, including the sensorimotor circuits for speech and its comprehension, and ancillary cognitive capacities for managing social and emotional interactions. It's hard to imagine such an integrated neurobehavioral system arising suddenly. Also a gradualist, Ray Jackendoff (2002) sketched out the likely properties of early protolanguage: symbol usage without syntax. Presumably, a syntactic stage would then have arisen later, perhaps 75,000–50,000 years ago, with a genetic basis as yet undefined.

Jackendoff's suggestion is, at least, plausible and consistent with several lines of evidence: in particular, the knowledge that human communication often involves several sensory channels (auditory, visual, and others) working in parallel—brings in the important subject of multimodality. For example, human speech is typically embedded in bouts of facial and gestural signaling. (Sit in any pub or coffee shop and watch two people in conversation.)

The other view positing a much more recent and abrupt origin for language is associated with linguist Noam Chomsky, famous for articulating the hypothesis of a "universal grammar." This is the claim that all languages share a common pre-existing design in our brains before we learn words; such unlearned core competence would then have enabled us to place words into meaningful phrases and sentences.

He and some others see recursion as perhaps the most interesting and most human aspect of language. Chomsky's account of language evolution is fundamentally about language as a process in our heads—as thought—decoupled from vocal performance and concerns about social interactions.

Either way, for success in social networks, humans clearly benefit from the ability to "do things with words." To take one dramatic example, anthropological studies have found that being a skilled storyteller—in a Filipino hunter-gatherer society—is associated with greater reproductive success (Smith et al. 2017).

Time and Imagination

A really broad view of language not only factors in multiple sensory channels, but also includes how we process narrative discourse, and this requires an explicit connection to memory. Memory will be considered in more detail in several chapters ahead, but we will touch on the related process of imagination here as it bears on language origins. Michael Corballis (2015, 2017) and Brian Boyd (2017) have also argued for a gradual evolution of language, in part because of its connection to gesturing and thus communicating in the visual domain. Briefly, their idea is that thought preceded language and that mimetic communication allowed us in a limited way to convey thoughts and intentions. Then with the development of episodic recall of past events, and imaginative anticipation of future ones, more complex forms of communication arose, using the vocal channel as well: "sharing our mental stories" (Corballis 2017: 120).

Interestingly, a slightly different version of gradual language evolution, also involving imaginative capacities, appeared at about the same time. Daniel Dor, in a book (2015) and several papers, has advanced the idea that language may have arisen from the advantages that accrued to early humans if they were able to speak of things outside the realm of immediate experience: transcending the present tense: "*this is my experience now.*" Or as Dor puts it:

> The communicator does not try to make some of his or her experience perceptibly present to the receiver. Instead, the communicator provides the receiver with a coded set of instructions for imagination, a structured list of the basic coordinates of the experience—which the receiver is then expected to use as a scaffold for experiential imagination: follow

the encoded instructions, raise past experiences from memory, and then reconstruct and recombine them to produce a new, imagined experience. (Dor 2017: 109)

In this account, "The listener is not invited to share an experience with the speaker. The listener is invited to create an independent experience, on the basis of the skeletal formulation of the received code" (109). In this way, we can speak across the gap between our private experiences, and with precision, about things that are remote in space and time. In effect, Dor believes language evolved as a social "technology" to instruct the imagination.

The models of both Corballis and Dor are very clearly not of the Chomsky type, because they are deeply communication-focused and social. They assume a more gradual evolution that was probably a natural follow-on to an earlier mimetic stage of communication (Donald 1991).

It would be hard to miss the parallels between this way of thinking about language, and the way Elaine Scarry describes the instructive mode by which authors construct vivid experiential and imaginative prompts within narratives (Chapter 2). Clearly, the strength of Corballis' and Dor's theories is that they embrace mechanisms tying language and cognition together at multiple scales (phylogenetic time, lifespan, and possibly also levels of neural processing—see below). There will be more on imagination in Chapters 6 and 9 ahead.

Words Migrating from Sound to Sight

We know with certainty that writing was invented many times (by the Sumerians, the Egyptians, and others). What were the prerequisites for such a leap? The acceptance of written symbols, whether etched in clay or on papyrus, required, according to Maryanne Wolf, "the sum of a series of cognitive and linguistic breakthroughs occurring alongside powerful cultural changes." Wolf refers to these as "epiphanies" (Wolf [2007] 2008: 25). A shift from words encountered first in aural modes and then visual modes also happens in individual human development. Linguist Ignatius Mattingly, using a biological metaphor, "famously proposed that 'reading is parasitic on speech.'" He was correct in the sense that a child's ability to learn to vocalize printed words "is critically dependent on their phonological, or speech sound, skills" (Hulme and Snowling 2014: 1). The significance of this statement will become clear below because processing the sounds within words, even during reading, remains one route to their recognition in the brain.

Words are especially intriguing to cognitive neuroscientists because in addition to sound they simultaneously convey meaning based on the way they are spelled: this is because of *morphemes*: structural units where changes or additions to words can occur—for instance, prefixes and suffixes—to which we must pay attention. Words can have multiple morphemes within them. Think of the English word "unknowable." It has three morphemes *un*-know-*able*. Each is created from letters and one or more phonemes (basic sounds); so words often have sub-lexical structures for sound and meaning. However, there needn't be correspondence between the number of morphemes and phonemes in a word. Unknowable has three grammatical morphemes, but four syllables, and seven basic sounds.

When analyzing how we read, it becomes obvious that the distinction between sounds and morphology is a brain-based distinction. There is more than one brain pathway for processing word-sounds, and word-images are handled by yet others. As cognitive neuroscientist Stanislas Dehaene observes, "All writing systems tend to jointly represent sound and meaning. It is as though the ancient scribes were aware that [our brain] projects [letter] shape information both toward ... regions coding for speech sounds and to ... regions coding for meaning" (Dehaene 2009: 176). His statement gives useful hints of what is to come below where the influence of culture will be explicitly considered.

Transitioning from words as sounds to words captured visually is also something that most of us experienced during development. Infants progress through speech proficiency in stages. First, in a *sensory phase*, they listen and begin discriminating between various speech sounds. That is, they begin in a phase of universal speech perception (birth to about six to seven months). After this, children enter a phase of language-specific speech perception when they strongly recognize phonetic components of their native language (e.g., see Kuhl 2010, 2015).

At about seven months, infants recognize recurring sounds and "babble": spontaneously producing speech-like sounds. Babbling becomes transformed into single-word utterances at about twelve months, which as indicated above can be propositions, followed in turn by the production of two-word utterances, which are more clearly so. A threshold is reached for the production of multiple word sentences, typically after twenty-four months (e.g., Aitchison 1998). This then segues into an accelerating process of vocabulary growth. Although children first acquire words as part of speech, a "visual word form" recognition process builds off that ('parasitizes' aural recognition competency) as reading emerges. When our minds grasp the notion of "words," vocabulary development proceeds with impressive speed: children have periods when they may learn thousands of words per year.

Reading development is more variable in its profile than speech development and it is heavily influenced by a specific cultural context—schooling. There are some milestones: one early accomplishment, common to speech and reading, is the ability to assign names to objects, and this dawns by about eighteen months—when infants have already been experimenting with words for some time. Neurologically, it indicates an ability to integrate information across the cognitive dimensions of language: recognizing recurring sounds, then associating them with objects or actions, identifying visual patterns of letters and words, and finally conceptual categorization. This all happens within the span of a few years.

For a perspective on how miraculous the attainment of reading is, consider that, "although it took our species roughly 2,000 years to make the cognitive breakthroughs necessary to learn to read with an alphabet, today our children have to reach those same insights about print in roughly 2,000 days" (Wolf 2008: 19). How is this possible? It is certainly facilitated by the fact that reading is "parasitic" on the early, genetically guided appearance of speech. A key question then becomes: if reading and writing are not genetic, how do they make their appearance? Dehaene and others have been working on how vision comes to play an important role in word recognition, and culture too plays an essential role. We will examine Dehaene's ideas in some detail after pausing to explain the sort of methods cognitive neuroscience has brought to answering fundamental questions about operations that enable us to do things with words.

Words in "Brain Space"

In groundbreaking work during the 1990s, neuroscientists Steven Petersen, Marcus Raichle, and Michael Posner used positron emission tomography (PET) scans to differentiate the brain activity underlying the processing of words as they are heard or seen. The results confirmed general expectations from neuroanatomy: different sensory regions appeared active depending on which sensory channel was used. Motor and "premotor" cortex were engaged for words uttered, whereas cognitive manipulations, for example, finding verbs that could "work" with specific nouns—ball with "throw" or "kick" but not with "stand"—caused activations far forward of auditory and visual cortex (in the temporal and occipital lobes, respectively) and into the frontal lobes (Raichle 1994, Petersen et al. 1998). Images from these experiments, all showing the left side of the brain, were widely reproduced, and were taken as a sign that, in the future, our understanding of language was going to be heavily influenced by brain imaging advances—something that has been borne out.

A bit later, Ksenija Marinkovic and her colleagues (2003) made the processes associated with language even more readily visible as they happen in real time. In her experiment, subjects performed a semantic judgment task when responding to words either heard or read: for example, they might be asked to judge the size of objects, animals, etc. denoted by words and respond to those larger than one foot (e.g., tiger), but not for smaller ones (e.g., cricket). At the same time, their brains were scanned with a special version of magnetoencephalography (MEG), which is basically a magnetic version of EEG (electroencephalogram, or "brain wave" recording). Given the high level time resolution of MEG, her results went beyond static PET images to produce dramatic "brain movies" showing the sequence of cortical areas handling words, sampled at a rate of about 200 times per second.[3] She has referred to this as "a word's voyage" through the brain (Marinkovic 2004: 143).

On trials where a word was heard, the scans recorded activation in the temporal lobes beginning about 150ms (milliseconds) after word presentation—this time is taken up by the ear responding mechanically, then neural signals moving along the auditory nerve, and through the brainstem before arriving at the cortex. Activation in primary auditory cortex (prominently on the left side) was followed by an expansion of activity into immediately surrounding auditory cortex. Activity peaked somewhere between 400 and 500ms and also spread forward in the temporal lobe, finally initiating an additional burst of activity in the ventral prefrontal cortex. Most of the neural activity was over by about 700ms.

On reading trials, initial activity arose in visual cortex (occipital lobe) about 100ms after word presentation (again prominent on the left), then it spread forward into the ventral occipital and caudal temporal lobes (peaking between 150–250ms). Similar to trials on which words were heard, the wave of activity spread forward in the temporal lobe, reaching the anterior temporal pole. Subsequently, activity also moved into ventral prefrontal areas. Most activity was over by about 550ms. So we can see in these "brain movies" that 500ms is at least a good ballpark estimate for the interval of brain time taken from detection of a word to performing a task, such as determining the word's meaning. These "brain movies" are a reminder of just how dynamic and distributed brain activity can be, even on exposure to a single word. The "tidal surge" of neuronal activity supporting word recognition is swift indeed.

Where (and How) Do We Recognize Words?

In Chapter 1, we mentioned that as a person reads, their eyes do not scan the page smoothly, but make saccadic jumps from word to word. This is part of the context for what Maryanne Wolf calls a word's "500 milliseconds of fame." Because of saccadic eye movements, each word, or segment of a larger word, is serially presented to the retinal fovea—the area of high-resolution vision. Then retinal signals pass via the optic nerves through the thalamus and up to visual cortex in the occipital lobe at the back of the brain. There are many ways to display the neural events involved in grasping a word. We will present a highly schematic model capturing only some key features.[4]

From basic neurophysiology, it is likely that orienting the eyes to a group of letters, and initiating letter and word form recognition, takes about

150ms. Full processing of word forms and component sounds takes perhaps another 200ms, and processing for meaning (semantics) adds about another 100–150ms or so. Therefore, processing a word through this channel is again about a 500ms cognitive operation.

Is discourse comprehension really as simple as a sequential flow of lexical information across several cortical areas, like watching a train of letters or word segments? Not at all. There is more going on than we show in Figure 5.2. As words or their components are processed visually, information flows to higher levels, so-called bottom-up processing, where meaning and understanding emerge. However, there also is much evidence

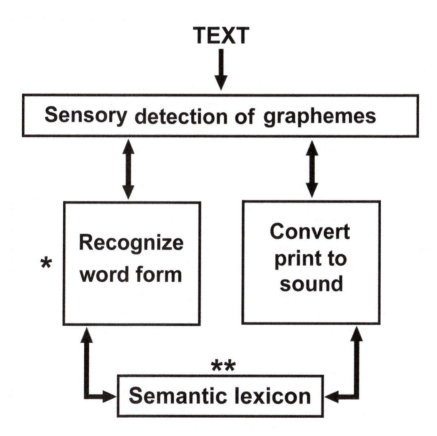

FIGURE 5.2 *A simplified cognitive model of some steps in word analysis during reading. This shows that analyzing letter shapes is a separable process from analyzing word sounds. Information would not likely flow in only one direction from the first box toward the lexicon, but also from the lexicons back toward early processing steps. The extent to which word forms need to be recognized first or in parallel with sounds is not fully resolved. The two processes identified with* **asterisks** *have been probed and localized in studies described in this and the next chapter.*

for "top-down" processing, where higher-level centers influence the earlier stages of reading—probably assisting rapid re-evaluation based on situational context. This is consistent with what we saw in Chapters 3 and 4 about sensory pathways being hierarchical, but also containing two-way flows of information, and parallel processing of information, together with lots of "flow management" by higher centers.

Although simplified, this scheme does allow one to grasp an essential set of processes. The visual shapes of letters must be processed as *graphemes,* abstract representations of letters, independent of typography. Think of all the different type fonts designed across history, and the variability of handwriting: we can read consistently despite such variability. Furthermore, graphemes are independent of orientation:

$$\text{"a G is a} \quad \cap \quad \text{is a } \mathcal{G}\text{"}$$

In one standard lexical route, a word's graphemic representation is first identified [* in Figure 5.2] and then eventually compared against stored representations of word meanings in long-term memory (a semantic *lexicon*) or dictionary leading to understanding. Some analysts have assumed we have a separate lexicon for word sounds (phonology) as well (not shown in the figure). Recent experiments to characterize our brain "dictionary" make this less likely (see Chapter 6 for a consideration of our semantic lexicon [** in Figure 5.2]). But consulting such maps of meaning, we may find, yet again, several possible routes to the same end.

We mentioned multiple pathways and parallel processing, and indeed these might be expected. Why? Well, because, they support each other for optimal word identification. For example, when reading an unfamiliar word, it appears there is a quick conversion of graphemes (print to sound). We all recall trying to pronounce unfamiliar words (out loud or sub-vocally) to make sense of them. However, if a word is frequently seen it may pass along the more direct route from visual grapheme to semantics. We can surmise that the various routes have different emphases during learning, when most words are necessarily unfamiliar.

It is worth pausing to appreciate the fact that, yet again, no amount of introspection would allow us to understand that this is how we actually begin doing things with words. And as Marinkovic has noted, "When we read a word or when we hear an utterance [in our native language], we derive its meaning effortlessly and automatically. In fact, we cannot choose *not to* understand a meaningful word that is communicated to us" (Marinkovic 2004: 147). Just as sensory processing is automatic, so too is semantic retrieval. As with so many cognitive processes, we know we can do it, but we don't know how. This is where the analytical techniques of neuroscience are invaluable: they make hidden processes manifest.

Minding Our "Ps" and "Qs"

So, what about word-form recognition, is it just a step within cognitive models? Where can we see it happening? Well, it is quite real and has been traced to a specific pathway and node as seen [*] in Figure 5.2. Cognitive neuroscientist Stanislas Dehaene's group has conducted years of work on these processes as they underlie reading (Dehaene 2009).

When letters are seen, information about them makes its way to the visual cortex at the back of the head. From there, it moves into the inferotemporal (IT) cortex. Here Dehaene's team found a distinct region that they believe provides the crucial next step—analyzing letters to enable the identification of words. They named it the *visual word form area* (VWFA—Figure 5.3).

Tying even one step of a neurocognitive process to a brain area is no easy matter, so the evidence must be carefully considered. Occasionally, people are found with lesions in this area, and while they can speak and write, or distinguish faces and objects visually, they cannot read words: this is a specific "word blindness," known to clinicians as *pure alexia*. Like many other aspects of the story of language, its antecedents go back more than 100 years: a patient with this condition, and left-side cortical damage, was first described by the neurologist Dejerine in 1892. However, the exact location of such disabling lesions was not pinpointed until recently, when Laurent Cohen and others analyzed several alexic patients (Cohen et al. 2003). They found that lesions producing the deficit were indeed associated with the brain region that Dehaene calls VWFA (Figure 5.3).

Other findings supporting a crucial role for this area in reading come from functional studies. To clarify, the VWFA is usually found only on the left side of the brain, so, while you will hear ahead that higher level processing for speech and reading is really bilateral, this early stage of processing is lateralized. Another relevant detail is that VWFA is especially dedicated to foveal vision (not all visual areas are so keenly dedicated in this way). Importantly, its neurons respond to letters in a manner that displays "spatial invariance." That is, they will respond to letters equally well even if their size, orientation, or font is altered.

We mention this invariance because this is exactly what's needed at a word recognition site in the brain because such invariance is characteristic of our reading behavior with its reliance on graphemes. Whether we see the letter R or *R* on a page, the names Nico or Ni*C*o, Luke or *Luke*, we can read them effortlessly. In short, the VWFA has an ideal neurophysiological profile to support reading as we perform it under real conditions.

Interestingly, there is another form of invariance that can present challenges for readers, and may shed light on the origins of the VWFA. In most routine visual recognition, we benefit from *not* bothering to distinguish between objects when they are rotated in a mirror-symmetrical way about their vertical axis: so-called mirror invariance. There is a good reason for

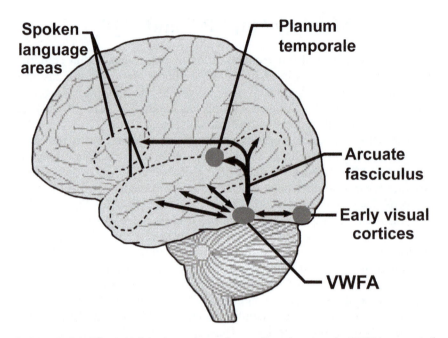

FIGURE 5.3 *The initial brain pathway for reading showing the VWFA as a vital node for recognizing words. The regions outlined by the dashed lines are involved in processing spoken language prior to reading acquisition. The visual word form area (VWFA), located in the left IT cortex, is thought to play a pivotal role in reading acquisition by enabling the rapid recognition of strings of letters and their translation into sequences of sounds. The regions in gray—the VWFA, early visual cortices (V1 and V2), and planum temporale—show enhanced neural activity with reading acquisition. The connections between them (shown by arrows) may also become enhanced— this has already been demonstrated in the case of the posterior part of the arcuate fasciculus (thick arrow). From: Dehaene et al. (2015) Illiterate to literate: Behavioural and cerebral changes induced by reading acquisition.* Nature Reviews of Neuroscience *16:234–44, Figure 1a, some details enhanced for clarity. Used with permission.*

this: in nature, objects such as predator and prey must be recognized quickly, no matter what angle they are seen from. But in the specialized context of letter identification, this can present problems: a letter like "**b**" when rotated around its vertical axis becomes "**d**," an entirely different letter.

Most literate adults do not confuse these two letters, but, significantly, when learning to read they did: children normally go through a stage where letters are misread, and especially if the letter in question is capable of mirror rotation. Indeed, at a certain point, most children print letters or words in a mirror-reversed way. This resolves itself with continued learning and is only a problem if it continues past age nine or so.

VWFA is nested within a region of cerebral cortex surrounded by areas visually sensitive to a range of other stimuli, such as faces and objects. These surrounding areas, in fact, do show mirror invariance in their responses to such visual categories. For example, a cup is recognized as a cup no matter whether its handle is to the left or if rotated around so that the handle is on the right. As a result, to create skilled readers, the area that will become VWFA "learns" to represent letters without mirror invariance. A "p" is processed differently than a "q" and we see them as different. This is not the case for people who have not learned to read! Their IT cortex, where VWFA should be located, displays mirror invariance for letters—like the mirror invariance for objects in the rest of IT cortex. Although it's hard for many of us to imagine, non-readers simply cannot tell the difference between "p" and "q."

Bringing together these observations, Dehaene proposed that spatial and mirror invariances are intrinsic to IT cortex, where they serve object recognition well. However, in learning to read, we selectively "suppress" mirror invariance in the specific region that will become the VWFA. This suppression shows up in most literate adults about 100ms into processing words (e.g., Pegado et al. 2011, Pegado et al. 2014), but it is undetectable in people who are not literate.

Dehaene and Cohen laid out their larger ideas on how the VWFA comes to support reading through a process of cultural reinvention (Dehaene and Cohen 2007). First, they dismissed the idea that we inherit a word processing "module" in our brains because evolution would not have had time to form one; writing systems are a recent invention. In addition, there is historical evidence that, over time, multiple writing systems with different characteristics were devised and, furthermore, throughout history only a small percentage of the human population ever possessed this skill (Wolf 2008).

So, Dehaene and colleagues came up with a specific, testable model whereby culture can be "written into" our brains: Their hypothesis asserts that the VWFA is a "domain-specific" structure, that is, one dedicated only to visual cognitive processing. They note that, while there are a cluster of different visual processing areas in the IT region of the primate brain, reading functions are inserted there selectively: VWFA is in roughly the same location in all literate humans.

Their striking conclusion then is that cultural activities are "inscribed" in our brains, but not in arbitrary ways, rather according to constraints based on inherited neural organization. More precisely, they have found that "domains of knowledge, such as reading, but also arithmetic, map onto remarkably invariant brain structures (often called 'cultural maps') with only small cross-cultural variations" (Dehaene and Cohen 2007: 384).

This raises the obvious question. Why couldn't we learn to decode word-forms almost anywhere in our visual brain? Dehaene suggests that before reading takes root in the VWFA, it is a region whose cells code for features of visual edges and their intersections—helping to define key features of objects—just as this area does in our non-reading primate relatives. However, some human cultures create pressure on individuals to co-opt it for encoding the line intersections within letters, such as F, L, T, Y. In essence, Dehaene's model hypothesizes that a region of primate IT cortex happened to be "pre-adapted" for a role in letter recognition because of its role in more generic object recognition (Tanaka 2003, Reddy and Kanwisher 2006, Sato et al. 2013). Using current terminology, one would say that the primate IT was "exapted" for human cultural purposes.[5]

Findings are coming together to flesh out this model for reading, and studies of the effect of literacy on the brain are proving insightful. Some idea of the developmental dynamics can be appreciated from Figure 5.4.

Children who are "good" readers show evidence on functional magnetic resonance imaging (fMRI) scans of a region selective for words in the left IT cortex (bright area identified with arrow in Figure 5.4)—this is Dehaene's VWFA, a region almost undetectable in those not literate. VWFA is immediately adjacent to areas responsive to pictures of faces, bodies, places, and tools, as documented by Dehaene's group and many others, and its presence can be detected within the first few months of life (Deen et al. 2017).

Response to written sentences

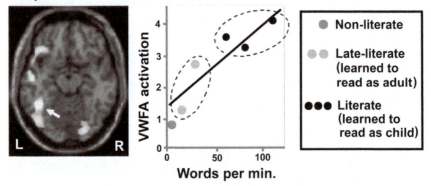

FIGURE 5.4 *Literacy is reflected by important changes in IT cortex (VWFA) visual responses. The brain section at **Left** indicates the cortical locations at which activation evoked by short written sentences is positively correlated with literacy level (measured by number of words read per minute). A particularly strong correlation is observed at the VWFA site (indicated by an arrow in the image, additional areas are activated by visual or sentence content). Data points on the **Right** represent the average activation of the VWFA in each of six groups of subjects with different levels of literacy (medium gray = non-literate people, light gray = late-literate people, and black = literate individuals. Within the dotted ellipses reading ability was greater in groups to the right). From: Dehaene et al. (2015) Illiterate to literate: Behavioural and cerebral changes induced by reading acquisition.* Nature Reviews of Neuroscience *16: 234–44, Figure 1b. Used with permission.*

Once some cells in this region start losing mirror invariance and sharpening their sensitivity to certain visual edges, the organization for reading capacity begins to take hold. Dehaene says:

> The acquisition of reading seemed to induce an important reorganization of the ventral visual pathway, which displaces the cortical responses to faces away from the left hemisphere and more toward the right. This displacement is presumably because the features that are most useful for letter recognition (configurations and intersections of lines) are incompatible with those that are useful for faces, so that one pushes the other away.
>
> (Dehaene 2013: 9)

Reduced capacity for face recognition on the left does not mean that face recognition is less efficient in literate people; their processing seems to be equivalent or even slightly more efficient than in non-literate people.

Finally, with regard to mirror invariance, "Its presence prior to literacy and its disappearance during reading acquisition implies that learning to read involves the recycling of a preexisting circuit that never evolved *for* reading but willy-nilly adapts to this novel task." The work on comparing literate and non-literate people has also revealed spin-off effects on speech perception: "[Their] brains differ in the very manner in which they encode speech sounds" (13). As a collateral effect, learning to read seems to increase the ease with which children comprehend complex *spoken* sentences (Dehaene 2013). The changes happen prominently not only in VWFA, but also in related parts of the reading circuitry (see Figure 5.3).

This is an amazingly rich body of scientific work with direct bearing on reading, much of narrative, and virtually all literature. Like most research, it has been scrutinized and debated. When these results first began to appear, some other researchers were skeptical that VWFA really existed, insisting on referring to it as the "putative VWFA" (Baker et al. 2007, Vogel et al. 2012). Skepticism, counter-proposals, rebuttal, corroboration: this is the way it's supposed to be. We advance from confusion toward less confusion (never perfect knowledge) by challenging findings and trying to replicate experiments. And keeping an open mind: it is quite possible that VWFA performs some tasks in addition to what has been attributed to it so far.

In an enlightening parallel, monkeys have a roughly similar organization to humans in IT cortex: they too have visual areas with selectivity for the basic categories of faces, objects, and locations. In tests based on Dehaene's work, neurobiologists Krishna Srihasam and Margaret Livingstone (2012, 2014) found they could train monkeys to discriminate human letter and number symbols visually—something they do not normally do. Monkeys trained with these symbols early in life, but not later, became "fluent" in discriminating them. Furthermore, the visual brain regions of these monkeys changed following training. The "trained" region was in the same cortical location in each monkey, in an area roughly equivalent to the human VWFA. So it appears early training can cause the formation of a specialized cortical region capable of processing letter-like symbols in other primates.

This sort of plasticity and learning in monkeys "is constrained by some *native* organization in cortex" (2014: 1782, our italics). This is analogous to the conclusion Dehaene's group came to in their human studies: culturally derived experience can shape neural function in the cortex, but there are inherent constraints. Finally, human cultural differences add another factor: one that shows a relationship between individuals and the social groups into which they were born. The VWFA in humans is reported to be more responsive to text characters in a subject's own language than to textual shapes in unfamiliar languages (Baker et al. 2007). Reading competence is based on cultural forces, with some individual linguistic variation as an overlay. Hence, some read easily, others not so much. Some derive great pleasure from it, others use only as needed.

Notes

1   The term "mytheme" was coined by anthropologist Clause-Levi-Strauss.
2   *FOXP2* codes for a "transcriptional factor," a controller of *gene networks*, meaning its activity regulates the expression of several other genes. As expected, it is expressed in human brain cells.
3   See the movie in the Supplemental Data section of their paper online at https://www.ncbi.nlm.nih.gov/pmc/articles/PMC3746792/
4   Our simple diagram is a synthesis of several models of reading, see for example, Coslett (2003); Hillis and Rapp (2004). For a developmental perspective, see Ramus (2004).
5   The term "exaptation" refers to a trait that originates performing one function and which is later co-opted for another function.

6

The Cognitive Habitat of Narratives

In a dramatic memoir, *The Diving Bell and the Butterfly*, Jean-Dominique Bauby ([1997 French] 1998 English) provides a moving account of someone who has lost all sensorimotor function yet retains language. At a relatively young age, and with a high-flying publishing career, Bauby suffered a massive stroke that left him with so-called locked-in syndrome. Unable to feed or breathe for himself, and with movement restricted to the blinking of an eye, he faced a grim future with little prospect of improvement.

Bauby's reaction to this nightmare was to engage with the world in the only means available, that flutter of an eyelid, and a specially designed alphabet, based around the most frequently used letters in French. Using these meager resources, he spelled out, letter by letter and word by word what he wanted to say. An assistant read the sequence of letters from the beginning, until he blinked, to signify "yes," then that letter was written down. The assistant would then re-start the process from the beginning, until, once more, the blink of an eye signified an affirmative. And so on. This arduous process went on for hours until a sentence, then a paragraph, and, miraculously, a page had been completed. The book's very existence is a testament to sheer, single-minded determination.

What is extraordinary, even beyond the laborious writing process, is the surprisingly tranquil, measured, and beautiful prose that resulted, lightened throughout with the dry humor Bauby brings to his retelling of what must have been a hellish existence. Here, he has just seen himself in a mirror for the first time since his stroke. "Not only was I exiled, paralyzed, mute, half deaf, deprived of all pleasures and reduced to a jelly-fish existence, but I was also horrible to behold. There comes a time when the heaping-up of calamities brings on an uncontrollable nervous laughter—when, after a final blow from fate, we decide to treat it all as a joke" (33). Only the subtle shift of personal pronoun here—"I" to "we"—betrays the discomfort of a man isolated in his own personal prison.

Some of Bauby's best moments came when he was wheeled onto a deserted terrace at the hospital, and his mind was suddenly unleashed. He dubbed this forlorn spot "Cinecitta," and, once there, gave full reign to his imagination, in this case fueled by memories of movies and extended fantasies of directorial indulgence. "I could spend whole days at Cinecitta. There, I am the greatest director of all time. On the town side, I reshoot close-ups for *Touch of Evil*. Down at the beach I rework dolly shots for *Stagecoach*" (37).

As he nears the end, his persistent inner voice and fleeting recollections are mostly what remain: with one eyelid sewn shut, suffering serious hearing disorders, he retains a heightened sensitivity to the fragile remnants of selfhood:

> When blessed silence returns, I can listen to the butterflies that flutter inside my head. To hear them, one must be calm and pay close attention, for their wing-beats are barely audible. Loud breathing is enough to drown them out. This is astonishing, my hearing does not improve, yet I hear them better and better. I must have the ear of a butterfly. (103)

Bauby, physically almost entirely incapacitated, retained a vibrant linguistic capacity—and consequently, memory, rich imagination, and an enduring sense of self.

Nothing could more clearly show that our language-related neural circuitry is not merely engaged in sensorimotor processing, but also with the cognitive substrates of memory, imagination, and emotion, which will become our focus from here on. In this chapter, we will examine some of these more extended linguistic functions using recent evidence from psychology and neuroscience, inquiring how much do we know about the processing of language beyond the recognition of words, textually or vocally? How extensive are the linguistic networks within our brain? How does meaning emerge from them? At last we are making some progress.

Sentences: Shakespearean Clues

With some fundamentals of word recognition established in Chapter 5, the more difficult question of how we comprehend sentences and then narratives is before us. One example of how to approach the question comes from a collaboration between a humanist and neuroscientists, combining the technicalities of observing brain activity with an appreciation of the playful possibilities of literary creativity.

Philip Davis, director of a center for reading and literature at Liverpool University, has described the strong subjective response to the language of Shakespeare, indicating that his lines seem to do "something almost directly physical within the human brain," or stated another way, Shakespeare can cause that "flash of lightening that makes for thinking (Davis 2007: 2, 93). This is redolent of Byatt's "electric" reaction to John Donne (see Introduction). Intrigued by this phenomenon, Davis collaborated with neuroscientists Neil Roberts, Guillaume Thierry, and others to explore what it was in Shakespeare's language that produced the "lightning." Together they focused on what seemed to be one likely cause: the playwright's deployment of "functional shifts," that is, the use of one part of speech to stand in for another. For example, a noun or adjective was sometimes used where a verb would be expected "strong wines thick my thoughts" or "the dancers foot it with grace."

Pursuing this lead, they randomly presented sentences for subjects to read that—by *changing only one word*—either:

1   were conventional in syntax and semantics,
2   contained a word-switch making the syntax and semantics odd,
3   were syntactically correct, but still semantically inappropriate,
4   involved a Shakespearean-style functional shift—syntactically
     odd—but nonetheless semantically graspable.

From simultaneously recorded electroencephalograms they found that the response to sentences of type 4, with a functional shift, displayed distinctive features: specifically, a voltage wave indicating a problem with syntax, but no trouble processing meaning (Thierry et al. 2008). For professor Davis, this was exciting—as he put it, "a chance to map something of what Shakespeare does to the mind at the level of the brain."

Crucially, the experiment proved to be "robust," that is, reproducible, so the group repeated it while scanning subjects with fMRI (Keidel et al. 2013). One of the interesting findings to emerge was that brain areas distinctively engaged by functional shifts were not located solely on the left side of the brain but were bilateral. As Davis says, "The concept of 'shift' extends to the domain of functional neuroanatomy, since it leads to a shift in activation from traditional [left hemisphere] structures to 'additional' [right hemisphere] networks usually involved in processing non-literal aspects of language" (917). A number of studies suggest that figurative expressions, new metaphors, and humor all activate the right side of the brain in frontal regions—but not those areas on the left (Bartolo et al. 2006, Mashal et al. 2007, Pobric et al. 2008, Marinkovic et al. 2011).

These experimenters also observed that functional shifts activated the basal ganglia on the left side of the brain. This is interesting because these structures have been implicated in handling unusual lexical processing, what we can't do automatically, and Shakespeare's wordplay stops us in our tracks. Consider an example of functional shift from *Othello* that Davis cited in his work: "To lip a wanton in a secure couch." As he says, "This line may serve two distinct masters. On the one hand, Iago is actively goading

Othello to commit a terrible act, and uses this vivid language to stoke the Moor's anger. On the other hand, Shakespeare wants to disturb the listener and violates linguistic expectations to achieve this end" (Keidel et al. 2013: 917). Such moments are not unique to reading *Othello*, or even Shakespeare, but rather are an example of the engagement of extended neural resources to handle literary devices that are unexpected, and challenge our automatic, habit-bound, comprehension.

Remember Lok's incomprehension facing a hostile human with a bow in *The Inheritors*? "Lok peered at ... the lump of bone and the small eyes in the bone things over the face. ... The dead tree by Lok's ear acquired a voice." Grammatical and semantic incongruity whether achieved by way of functional shift, analogy, point of view, or metaphor may add to the "literariness" of how we read in part by altering attention during sentence processing.

It is intriguing in the Davis study that, in addition to areas of the temporal and frontal lobes already known to be activated by words, there were areas of *cingulate cortex*, the *precuneus*, and the *dorsomedial prefrontal cortex* activated by Shakespearean functional shift. These are areas of the cortex not classically thought of as linguistic, but recently associated with the processing of sentences or narratives, and with connections to emotional and introspective aspects of thought. All three structures are part of the so-called default mode network or DMN, one of the recently identified cognitive networks mentioned in Chapter 3. Davis' findings encourage an examination of the DMN, and we will provide more information about it below.

The Dictionary inside Your Head

When a word is encountered in the context of a sentence, its semantic and syntactic significance must both be processed: the specific regions of the cerebral lobes where these two operations take place at least partly known.[1] For example, both processes involve similar regions in the parietal lobes, and adjacent but distinct parts of the frontal lobes (in Broca's area); semantic processing *also* involves significant regions of the anterior temporal lobes (ATL). So both linguistic functions involve areas of several lobes with a degree of overlap (Friederici et al. 2000, Hickok and Poeppel 2007).

Linguists have long believed that we understand a word by matching its neural representation with word meanings that we retain in a long-term cognitive storehouse, called a semantic *lexicon* or, simply, semantic memory [** in Figure 5.2]. It is difficult to segregate memory, and other cognitive functions such as imagination and emotion, from language. However, we know that an essential process to understand a word is to scan semantic memory leading to the recognition of its meaning(s), and this must happen iteratively as we listen or read. Cognitive science has specific ways of describing what may be going on and a good description has been provided by Patrick Colm Hogan (Hogan 2003b). In this chapter, we construct a broad overview of some recent neurobiological findings about semantics.

Long-term memory is not a unitary system, but rather a highly differentiated set of systems (Chapter 9). Two we know well from personal experience are a system of linguistic (and also factual) knowledge about the world, *semantic memory*, and personal recollections of our life's episodes, *autobiographical memory*. Semantic or language-based knowledge is an area where impressive progress has been made in recent experiments.

Any taxonomy of knowledge recognizes that we know things about a series of *domains*, such as people, other living things, objects, places, etc. The importance of this for our language fluency and our ability to fashion and tell stories is hard to overestimate. To take just one example, making connections across categories is basic to metaphor.[2] As to such links between our organizational domains and cognition, Ray Jackendoff has pointed out, "We can understand things in the world as belonging to categories only because we (or our minds) construct the categories" (2012a: 132). Indeed, we can say a number of specific things about the way our brains do this very categorization.

When we "consult" our lexicon, we bring word information into a functional entity called *working memory*, where we can use it in various ways. For example, studies by psychologists have repeatedly demonstrated that there are functional relationships between items within a domain. If you have recently accessed a word relating to family relationships, then you will typically have a brief period where you can access other words about family more easily than words in other domains, such as tools. This is called *priming*. The very existence of such category-specific priming suggests that there is a neurological reality to the idea of semantic domains.

Semantic knowledge can be probed with words directly, or by the use of pictures to represent items in different categories. Remember that, as visual information moves beyond initial processing, it enters a ventral stream where object identification predominates (Chapter 4). We also saw in Chapter 5 that the VWFA is situated in a ventral area that also contains zones with neurons displaying visual selectivity for objects, faces, and places. This ventral (or "what") stream continues anteriorly into the temporal lobe, and into lower parts of the parietal lobe (in Figure 5.3, this was shown by arrows moving anterior and dorsal from VWFA). Ultimately, there are also connections with the frontal lobe; all of this neural terrain has yielded to increasingly detailed analysis in the past few decades.

These linguistically significant regions are interconnected by bundles of nerve fibers running through the white matter under the cortical surface, so they are in communication via an extended "transport network" (Bajada et al. 2015) during semantic and related lexical operations. The effects of damage within these areas suggest that linguistic comprehension occurs here. Lesions at many places in this network affect not only semantics but sometimes also syntax, for instance, producing difficulties understanding spoken words, or assigning names to objects presented visually.

A comprehensive discussion of syntax and semantics cannot be undertaken here. Yet in the light of highly promising findings in neuroscience, we will touch on the issue of how meaning and knowledge are represented in the brain. So we begin: wide damage to parts of the extended semantic system just described above can be devastating, for example, bilateral atrophy of the ATL produces profound "semantic dementia." Following up these clinical observations, recording and imaging studies have shown that various degrees of specialization for particular categories of knowledge are displayed by neurons along the ventral stream and in the temporal lobes (Martin et al. 1996, Cummings et al. 2006, Pulvermüller et al. 2009, Ungerleider and Bell 2011, and see review by Jeffries 2013). Neuropsychologists have also documented how small lesions there can lead to loss of the ability to comprehend or name items within only one category (such as living things or man-made objects). A small hole in your brain creates a small hole in your knowledge. Thus there is abundant evidence for the semantic network displaying "category specificity," although exactly how many categories exist within our lexicon is unclear (but see Figure 6.1 and related discussion for some possibilities).

In the last few years, ever more refined techniques, operating at increasingly detailed levels, have enabled us to observe word recognition and access to semantic "entries" in the neural "atlas" (where you maintain words currently in semantic memory). Several general conclusions have emerged. One is that beyond the ATL, a rather large area of cerebral cortex may be involved, beyond those just mentioned, but the specificities of regional involvement have just begun to be worked out.

That being said, we know some semantic zones show specificity that is modality-related. For instance, words connected to visual properties may be handled adjacent to visual sensory areas, those for acoustic properties around the auditory cortex, action referents may cluster near cortical motor areas, etc. This has been generally known for some years now, but even more specific information is emerging. For instance, a number of studies have uncovered evidence for semantic activity within "supramodal" cortical "hubs" where several senses converge, and it seems that such sites are specifically connected to words with more abstract meanings (e.g., Binder and Desai 2011, Pulvermüller 2013).

So an emerging model is that "hubs" (in ATL and other cortical areas) are where significant amounts of semantic access occur, because they connect to a range of specialized sensorimotor areas through neural connections called "spokes." Thus the particular spokes of a hub determine what sorts of perceptual, action, or emotional meanings can be retrieved there (e.g., Pobric et al. 2010, Jeffries 2013, Chiou et al. 2018). Such spoke-and-hub models (similar to airline route maps) were mentioned in Chapter 3 in relation to large-scale brain networks. Their extensive use for our brain's semantic system is telling for, as you will see shortly, our semantic "atlas" seems to reside within a very large-scale network. There is now a body of experimental evidence for a good fit between our semantic system and a hub-spoke paradigm.

As indicated above, one area frequently regarded as a hub is the ATL. Bilateral damage to this area produces semantic dementia. But what evidence of its operation might be seen in the intact brain? First, functional brain scans (fMRI) show ATL active in situations where word meanings are accessed, and additionally and significantly, selective transcranial magnetic stimulation (TMS), a non-invasive way of temporarily disrupting function in a selected region of brain, can prevent retrieval of meaning (Jeffries 2013). This suggests hub-like function in the anterior temporal lobe. Sometimes it has been observed on the left side, but in other cases on the right side as well (Pobric et al. 2010). Other areas that some believe are hubs include anterior cingulate cortex (Zhao et al. 2017, see Figure 7.2 for cingulate cortex), and the middle fusiform gyrus—an area of the temporal lobe on the underside or lower surface of the brain—where both auditory and visual naming are disrupted by local intracranial electrical stimulation during surgery (Forseth et al. 2018). Importantly, these findings go beyond general correlations of brain activity with behavior and extend to direct testing of involvement using temporary, reversible lesions.

Semantics Embodied?

An additional thread (almost wrote "current": priming at work) running through most studies is the degree to which semantic cognition may be embodied. There are many findings but no clear consensus on this yet. Opinions range from "strong embodiment," where the retrieval of meaning requires processing in sensory and motor areas used during perception and action control, to models where semantic associations are quite abstract and unconnected with specific sensorimotor zones. Certainly, there are data that can be interpreted to support embodied semantics, but such results have not been consistently observed (e.g., Arevalo et al. 2012).

It seems likely, given the vast range of contexts in which we extract the meanings of words and phrases, and the potential for ambiguities, that meaning retrieval/understanding is facilitated in some cases by "extralinguistic" referencing to perceptual, action, and emotional circuits, but that it is not invariably required. Indeed, there is evidence that the nature of the task presented, and contextual features, influences which neural areas will participate in a particular act of linguistic comprehension, although the level contextual influences is not yet fully clear (e.g., Davey 2015, Lin et al. 2018, Stokes et al. 2019).

Atlas of Meaning

With regard to semantic studies, there is a trend within neuroscience that must be highlighted: a move toward ever more "ecological" (natural, not reduced) stimulus situations. So, for instance, instead of presenting experimental subjects with individual words or pictures and following neural processing while they identify them, make matches, draw distinctions, etc.—a reductive model of language use—people can now be studied while they process a full sentence, sets of sentences, or best of all, entire narratives, usually listening to them or watching a film. This requires significant computing power to collect and store vast amounts of fMRI data and correlate changes in the activity profile of all voxels (the tiniest brain volume resolvable in a brain scan) over the duration of the story.

A great example of this approach from research on semantics comes from Alexander Huth, Jack Gallant, and colleagues, mapping the brains of people with fMRI as they sat in a scanner listening to radio podcasts, entire stories 10–15 minutes in length. Their analysis produced a comprehensive model of activity across the cortex, time-linked to the processing of specific words. They used the enormous volume of data from each participant to see if there were regions where words with similar meanings were typically processed. Indeed, they found about 190 functionally distinct areas in the left cerebral hemisphere of which 77 were "semantically selective"—for example, some areas responded more for color words, some for words concerned with social relationships, others for words related to spatial navigation, etc. They also located about 130 distinct areas in the right cerebral hemisphere of which 63 were semantically selective. By combining information from all the participants, they were even able to create a "brain atlas" of the entire semantic system as revealed so far by their methods (Huth et al. 2016).[3]

The results in Figure 6.1 show only two sample areas, each with a "word cloud" indicating words that reliably caused activity to increase in the volume making up each distinguishable cortical area. While the details here differ a little from other semantic "maps" (e.g., Binder and Desai 2011, Pulvermüller 2013, Moseley and Pulvermüller 2014), there are important points of convergence. First, large regions of both left and right hemispheres are involved. It's only when we move away from initial linguistic processing and look at cognitive events that support narrative understanding that we see RH involvement more clearly.

Second, Huth and colleagues found that the data in their study could be naturally organized (using statistical methods) into twelve categories: some were *sensory* related, for example, visual and tactile, there were *numeric* and *temporal* categories, several *social* or *emotional* categories, and also an *abstract* category. Finally, there was a relationship between many of these categories and previously known features of cortical organization (similar to what we mentioned above from earlier studies). So, for instance, color words were found near cortical area "V4," which is known to compute color information for visual scenes.

As a consequence of all this, and the complexity of the real narratives used, individual words could be found at more than one location. If you were to ask where the word "family" is located, the answer is: several places, depending on context. Sometimes it occurs in the vicinity of words about social relationships, but also in areas activated by words related to living spaces. This echoes our everyday distinction between *denotation* and *connotation*. Given that words are learned, and heavily cultural in origin and meaning, it was more than a little surprising that, across all of the brains analyzed, the location where particular categories were found turned out to be similar. This strongly suggests that words become anchored in cortical

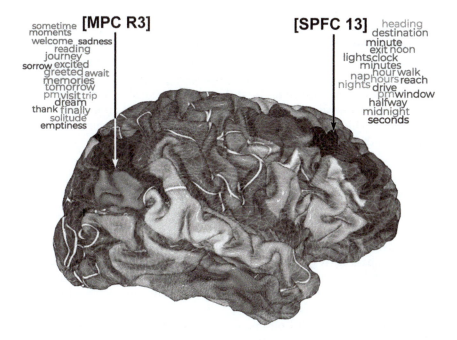

FIGURE 6.1 *Experimental results from the search for our semantic lexicon. This shows the lateral surface of the right cerebral hemisphere. At the front (black arrow), we show a region of prefrontal cortex that responded to words describing the natural world in terms of locations, time, and navigation. Toward the rear, we show a region of parietal lobe responding to words with a mixture of social/emotional terms but also some related to time. The gray arrow becomes white where it passes out of view between the two hemispheres to denote the position of this region on the medial surface of the hemisphere. Adapted from the PrAGMATIC Brain Atlas on the Gallant Lab website https://gallant-lab.org/huth2016/ (last accessed November 22, 2019). Material covered by creative commons attribution license. For related publication, see Huth, de Heer et al. (2016) Natural speech reveals the semantic maps that tile human cerebral cortex. Nature 532:453–8.*

"neighborhoods" that are functionally related, and that both biological and cultural factors probably shape features of this atlas of meanings.

This study, along with many others, upends the oversimplification that language is a left-brain function alone. That does not change the fact that the "final common path" for speech is located in the left hemisphere. But language is more than just speech; whether internal or external, it is part of thought, conceptualization, imagination, and our sense of self. In a recent follow-up study using the same semantic analysis approach it was observed that when reading a narrative is compared with listening to it, the same cortical regions are engaged, that is the "maps" so derived are highly similar (Deniz et al. 2019). So, despite arriving via different sensory inputs, from a semantic-cognitive perspective these modes of narrative consumption are treated the same.

It is worth pausing here to appreciate the fact that, once again, no amount of introspection would reveal this sort of amazing fact about our brains and the organization of meanings. Our cortex, whatever else it may be, is also a massive "semantic engine." As so often when studying cognition, we are examining phenomena that are largely preconscious or unconscious, and amazingly rapid. Not only is sensory processing automatic, so too are some of the processes that we think of as "higher functions"—such as semantic retrieval. As with many processes and behaviors, we *can* do it but we don't know *how*. This is where the analytical techniques of neuroscience are helping, rendering the latent manifest.

Behavioral Insights into Narrative Imagination

Questions of how we comprehend and respond to narratives and how those responses activate or intersect with memory, imagination, emotion, and attitude have been the subject of intense scrutiny by psychologists and narrative theorists for some time. But these dynamics have long been familiar to literary critics.

Consider Stanley Fish's outline of what we do when proceeding along a line of narrative: "The questions we ask in our reading experience are in large part the questions we ask in our day-to-day experience. Where are we, what are the physical components of our surroundings, what time is it?" (Fish 1967: 22). By considering how readers work through such questions to handle paradoxical statements and mental images (in his example, comprehending similes from *Paradise Lost*), he shows how real cognitive work is required (Fish 1967). In what lies ahead, you will see how psychology and neuroscience have tried to capture some of this narrative imagination in action. One of the outcomes is a view that aligns with Fish's that these cognitive operations of reading may be highly literary, but they are also related to our day-to-day needs.

As Patrick Colm Hogan pointed out in his handbook on cognitive approaches to literature (2003b), cognitive science involves the systematic creation and testing of models of intelligent behavior and its underlying processes. These schemes are sometimes general process flowcharts, but can be based on hypothetical circuits, or modules in the brain. Neuroscience really enters the fray when we know that proposed processes or modules actually exist—locating them in real brains.

In his meta-analysis of narrative psychology, Raymond Mar (2004) noted that cognitive theories of discourse comprehension advanced with some speed from the 1970s onward, but it took a while for neurobiological analyses to catch up. The focus in this chapter is primarily in taking the enquiry up to the point where evidence emerges about specific brain regions and their operations in processing narrative discourse. We do not consider important related questions such as the neural bases of aesthetic judgment (e.g., Massaro et al. 2012, Siri et al. 2018). For a recent example of this, see G. Gabrielle Starr's review of her research program and its major findings (Starr 2013).

Most accounts of narrative comprehension converge on some idea of what knowledge we take from texts or discourse. Some have suggested that what we store in memory is in a propositional form of narrative (van Dijk and Kintsch 1983) but others (e.g., Johnson-Laird 1983) have argued we store a "mental model," that is, an analogue of story elements, with a spatial character (see a good summary of these contrasting views in Emmott ([1997] 1999: 43–52). Such mental models are referred to by some as "situation models." By the end of the twentieth century, Zwaan and Radvansky (1998) proposed that the rise of these models suggests that, instead of just considering language semantically and syntactically in relation to memory, the text represents "a set of processing instructions of how to construct a mental representation of the described situation" (162).

As Zwaan and Radvansky stress, situation models are distinct from the concept of pre-existing schemata, traceable back to both I. A. Richards and F. C. Bartlett, which contain real-world knowledge we hold in memory—sometimes in a form called "scripts," for dealing with typical life situations (e.g., Schank 1995). These schemata can be thought of as culturally inflected expectations around recognizable events such as browsing a bookstore, visiting a doctor's office, or dining in a restaurant. Richard Gerrig (1993) using the example of a restaurant episode from an Ian Fleming novel suggests that these schemata enable authors to count on readers having basic information, which allows them to rapidly fill in textual gaps with details that support, and confer verisimilitude on, portrayed events.

What evidence exists for schemata and situation models? One clear example was presented by Bower and Morrow (1990) showing responses when we engage with a story through either reading or listening. Their interests were squarely on cognitive operations, which would include word recognition, acquisition of meaning, holding items in short-term memory, inference, and cognitive metaphors. Furthermore, they wanted to know how we identify *characters, causes, and actions* in a story, how we find the key agents, determine the sequence of their actions, and decipher causal relations. Such a mental model, once formed in a listener or reader, would be modified as additional statements and situations are encountered.

Besides paying attention to characters and actions, Bower and Morrow found that readers show a strong tendency to construct a map of the physical setting for a story. With respect to events, readers behave as if they locate characters within the setting, to render a situation map, which seems to help in creating a network of causal connections. (See Ryan et al. 2016 for a thorough and nuanced description of setting and space distinctions in the text and in the reader.)

Some narrative events are, of course, more significant than others— so one can think of a situation model as containing a primary chain of occurrences alongside others that are peripheral and weakly connected. We infer primary causation set against a character's goals, and this determines what will be retained in the reader's memory.

Yet stories are subject to the same limitations as other real-life experiences: what we recall is often not granular detail but the gist of what happened, and this central causal chain comprises just that. When psychologists examine narratives, they often look for their "coherency," a coherent story being one where causes and enabling events form a logical web of connections with outcomes. As proposed by Johnson-Laird, a text or description is coherent only if it enables a reader or listener to construct at least one internally consistent mental model of the situation (Bower and Morrow 1990). Of course, it's important to note that literary narratives need not be consistent in any logical sense. Ambivalence, ambiguity, and those functional "lightning flashes" are often to the fore.

Let's briefly examine the supporting evidence for situation models as having a spatial character. This is connected to how readers attend to fictional characters. In experiments where subjects had become familiar with the layout of a fictional set of rooms, they were occasionally prompted to make simple discriminations about objects or characters—such as decide if two objects are in the same or different rooms. Objects near the protagonist were discriminated more quickly than those near peripheral characters. This is presumed to be because those items are somehow more active in memory (or imagination) and so are easier to recall: another example of "priming effects."

Using this approach it has been shown that as a protagonist moves through a model space, the focus of heightened or primed responses moves with them. For example, when a protagonist is situated in a hospital waiting room we may be hypersensitive to mundane signifiers of time, or other patients' relative demeanor; once in an examination room, we are more predisposed to notice and remember other emotionally charged features— clinical instruments, a chart for scoring pain levels. In a particularly interesting finding, researchers looked at fictional examples when the protagonist was thinking about a place outside their "present" physical location—daydreaming, or projecting. In that case, reader's answer times were quicker for questions about objects near the protagonist's *mental* location, rather than their "actual" location in the story.

So, if the character was sitting in a waiting room but dreading the examination room, questions about instruments, etc. in the latter would typically be what "came to mind" more immediately for the readers. All of the above confirm the sense we have as consumers of narratives that we adopt the mental perspective of the central character—and further evidence confirmed that such models can also include representations of emotional state (e.g., Gernsbacher et al. 1992). So it seems fair to say situation models may display a complex and "map-like" quality.

A useful analogy is this: "The bare text is somewhat like a play script that the reader uses like a theater director to construct in imagination a full stage production" (Bower and Morrow 1990: 44). As a consequence, *it is your imaginative construct—your situation model—that has cognitive and emotional impact and is remembered rather than the raw text itself.* Bower and Morrow claim that the principles by which people understand actions in a story world are the same as in real life (recall Stanley Fish's similar point). This is a recurring theme emerging from cognitive research on narrative imagination. Given that memory and imagination both seem to figure in mental models of narratives, it may be helpful to show an explicit model of how this might happen (Figure 6.2).

In Figure 6.2, we bring together several of the processes that must be occurring as a situation model is formed, using input from ongoing perception and memory. The creation of a mental model, to the right, depends upon the work of the imagination and usually happens implicitly. The double-headed arrow indicates that interactions with memory (such as existing schemata or scripts in long-term memory, and new narrative items held in short-term memory) are means of updating such models.

There are key metaphors associated with narrative response that also demand our attention. One of the best-known is the idea of a reader being "transported" by a story: absorbed, disengaged from the immediate surroundings. Then we should ask where are they transported to? The obvious answer is to a narrative or story world: the phenomenological equivalent of a mental/situation model. Yet we should perhaps take a pause. As Gerrig has pointed out, although "narrative world and situation model circumscribe similar theoretical claims," the concept of narrative world is connected to broader theoretical aims that really consider "the diverse consequences of constructing situation models" (1993: 6, and see Herman 2013 for a sample of some of those aims).

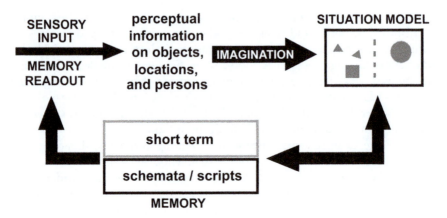

FIGURE 6.2 *Diagram highlighting involvement of memory and imagination in construction of a situation model during reading. The arrows connecting to memory indicate probable frequent interactions between memory and the situation model: providing schematic information during construction (left lower arrow); or storing the model and updating it (lower right, bidirectional arrow). Imagination is intentionally shown as a process rather than a location. While the situation model is shown separately for convenience, it might be equivalent to some aspects of short-term memory during reading.*

A related question suggests itself: what is the distance from the real world to the narrative one to which we are transported? Some commentators, following the "principle of minimal departure" (Ryan 1980), claim a storyworld is construed "as being the closest possible to the reality we know" (403), and while "great artistry might facilitate the journey, ... the only a priori requirement for a means of transportation is that it serve as an invitation to the traveller to abandon the here and now" (Gerrig 1993: 12). And we would add it seems to require the reader to engage "the work of the imagination" (after Harris 2000).

If we accept the notion that a fictional text may be a kind of imaginative "play script," then it follows that readers somehow "perform" the narrative. There are several corollaries, the most obvious being that reader response is not passive, but active—it is an act of imaginative construction. "Readers must contribute to their own experiences of narrative worlds. The conclusion is a general one: whenever we attend a movie, watch television, or read a newspaper, we are actively supplementing the 'text.' The same rule holds even when we are viewing paintings" (Gerrig 1993: 29). This performative contribution from the reader is strikingly similar to the concept of the "beholder's share" described by neuroscientist Eric Kandel with respect to visual art (Kandel 2012).

A nuanced "participatory perspective" has emerged in Richard Gerrig's work bridging cognitive psychology and narratology (Gerrig and Jacovina 2009). He has proposed a mode of "participation" that occurs at least partly outside of our awareness, meaning at least some of our imaginative work is implicit, whereas other aspects are explicit (conscious). Significantly, such "division of labor" is well-known in memory and might even have been predicted for imagination given recently discovered relationships between imagination and recollection at the neurological level (Chapter 9).

Linking Situation Models with Neuroscience

While a number of specific types of situation models have been advanced based on careful empirical analysis of reading, speech processing, picture processing, and more, we'll focus on a few contributors who helped establish attributes of such cognitive models and show how these have been analyzed within a progressively more neuroscientific framework.

In the 1990s, Morton Ann Gernsbacher began formulating a straightforward prototype of discourse comprehension based on reaction-time studies (Gernsbacher 1997). She observed that early in reading comprehension, we create a foundational model of some sort, cued by information in the text—often the first sentence. Subsequently, salient information is added and integrated, however when unrelated information is given, readers may formulate a substructure or an entirely new model. These model structures were hypothesized as nodes in a memory network, nodes here meaning model neurons linked functionally, thus Gernsbacher was relating her emerging construct to a cognitive process for which there was an abundance of data. Very little study had been conducted on processes of imagination at that time. An important feature of her model was the consideration of interactions between nodes based on enhancement and suppression, roughly analogous to synaptic excitation and inhibition (Chapter 3). Her work recognizes that we often speak of what is activated in the brain and forget that stimulus situations, external or internal, can trigger excitatory processes, inhibitory processes, or both. Gernsbacher used her approach to contribute to a study of dementia, showing there are detectable changes in discourse comprehension that result from failure of suppressive processes (Faust et al. 1997).

Then, in collaboration with Mark Beeman and Edward Bowden, she examined how each hemisphere's ability to make inferences contributes to story processing in relation to visually cued words. Words were projected either to the left visual field, seen only by right hemisphere (RH), or to the right visual field, seen only by the left hemisphere (LH) (see Figure 4.6 and related discussion). The experiment asked whether these words could be used in priming, for example, shortened time to recognition. An important feature of the experiment was that the words were presented either when so-called *coherence inferences* were possible, filling in information, or *predictive inferences*, implying future actions.

The outcome was striking. Early in reading the story, predictive inferences were generated for words shown to the RH; later in the reading process, coherence inferences were evoked by words shown to the LH. Importantly, when a story approached resolution both hemispheres contributed (Beeman et al. 2000). This shows that the "work" of drawing inferences is not strictly lateralized. The two hemispheres contribute at different times and in complementary ways: full inferential power requires both. However, there may be instances in which the inferential "style" of each hemisphere is advantageous: For example, Shears et al. (2008) followed up on differential inference formation by each hemisphere to show that the LH is better at working with concepts closely spaced in discourse and related to physical cause and effect, whereas the RH handles more distantly related concepts and this enables flexibility in predictions needed for planning. Thus, the RH is often characterized as having "coarser" semantics than the LH. So each hemisphere has greater abilities to work with distinct knowledge domains related to their differing inferential styles.

In a slightly different take on mental models, Zwaan and Radvansky (1998) proposed one based on "event indexing." Their suggestion is that a situation model of narrative should consist of related events: units based upon time, space, causation, motivation, character actions, and perspectives. Indexing is a way of mentally tagging events by relevance. Extremely relevant events, say, to the goals of a protagonist, are believed to be "foregrounded," that is, made more available in memory for activation. Take the example of going on a hike: it has a beginning and a destination. Suppose there is a pub with a roaring fire and lovely view at the top of the hill, entering the pub would mark a major event boundary. But there are often a number of minor event boundaries within any event, say stopping to identify flowers along the path of the hike, or scooping water from a stream to drink.

This type of model depends upon the prevalence of clear event recognition, the ability to divide a piece of narrative into meaningful segments. Helpfully, a group of psychologists and neuroscientists were working around the same time on the very topic of how we break up what seems like a continuous stream of sensory input into a series of events. They started looking at the segmentation of everyday visual events as portrayed in film clips (Zacks et al. 2001). What they found was a small surge of neural activity (with functional magnetic resonance imaging, fMRI) in very circumscribed regions of the cortex at points where participants indicated that they recognized boundaries. These areas were around the occipito-parietal junction bilaterally (area V5, which is sensitive to visual motion) and the right frontal cortex in an area associated with shifts in attention.

Jeffrey Zacks then expanded this work in a direction consistent with the idea and generality of event indexing. He established, also with fMRI, that as we read there are increases in activity in a few brain locations at key moments in narratives: moments that the readers had, in separate tests, judged as boundaries between events (Speer et al. 2007). Each clause in the narrative had been coded for changes similar to those in Zwaan and Radvansky's model: time, space, causation, objects, goals, and characters. All of these contributed to the recognition of event boundaries by participants. When correlations were sought statistically, transient activity increases were ultimately detected in more than half a dozen areas of the brain closely synchronized with the event boundaries (these areas were found in both hemispheres in the posterior cingulate, precuneus, and regions of the temporal and frontal lobes). This and more recent work (Baldassano et al. 2018) suggest overlap with parts of the DMN, see below).

This research has shown "not only that readers are able to identify the structure of narrated activity, but also that this process of segmenting continuous text into discrete events occurs during normal reading" (Speer et al. 2007: 454). Such results, it turns out, also apply when people view films (see Zacks et al. 2009). One interpretation of the functional significance of such neural marking of events is that it may reflect repeated active revision to bring the reader's situation model into line with new information, as predicted by these cognitive models (e.g., Figure 6.2).

What More Can Neuroscience Bring to the Table?

In 2001, Young and Saver published an analysis of studies conducted on patients with variously located brain lesions in an attempt to define the network we use in constructing narratives. They characterized several types of narrative impairments, "dysnarrativia," including two types of amnesiac alterations that followed from damage to the amygdalo-hippocampal system: narrative attempts that cannot extend past the time of brain injury ("arrested narration") and significant confabulation (or "unbounded narration") with frontal lobe involvement. Patients with specific ventromedial frontal lobe damage may display "undernarration": trouble inhibiting impulsive choices so they "run with" the first narrative option that comes to mind.

Most dramatically, people with damage to the more dorsal and lateral frontal lobes in both hemispheres displayed "denarrated lives"—they were aware of the world, but apathetic and unable to tell a story about themselves—indeed they seem to lack selfhood. Dysnarrativia also can result from generalized brain trauma, presumably by damaging some of the above-mentioned regions, so it is clear from recent data that both cortical hemispheres are involved in production and comprehension of narrative discourse (e.g., Sherratt and Bryan 2012, Karadamun et al. 2017).

Raymond Mar, in a 2004 meta-analysis of a number of studies using brain scans (both PET and fMRI) during comprehension of "narratives" (mostly just sequences of sentences), was able to detect consistent patterns of brain involvement in comprehension. Prefrontal cortex was widely reported to be involved, as well as mid-temporal and superior temporal cortex, the temporo-parietal junction, and posterior cingulate cortex. Across all studies, there were associations found for these structures on both the right and left sides, the details depending on what type of sentences were the focus (simple physical stories versus those involving mental attributions to characters).

Mar also analyzed studies of deficits following brain lesions, and reported roughly similar involvement of these areas, with more dependence overall on the RH, and the anterior temporal lobe (ATL) region. Finally, he looked at studies of deficits in narrative production in patients with lesions. His chief conclusion was that comprehension and production are not associated with completely separate brain areas. Again here, lesions to RH not just LH areas made a difference. Such lesion studies can tell us what sites are probably necessary for a cognitive ability, whereas brain scans may only show areas that are involved. Yet, to be clear, the participation of these areas is unlikely to be unique to narrative—some undoubtedly reflect general cognitive processes like selection of items, transactions with memory, drawing inferences, or detecting the mental states of others.

So, remarkable though this is, has neuroscience more to offer? Indeed, and perhaps the most exciting new research has moved toward using more "ecological" experimental situations. For example, Stephens et al. (2010) compared the neural profiles of storytellers with their listeners.

The brains of storytellers were scanned as they related a 15-minute unrehearsed real-life story which was recorded. Then the listeners were scanned while the recording was played. With a slight and predictable delay, the listeners' brains displayed activity patterns that mimed the speakers' brain activity. What is really fascinating is that separate tests revealed a strong quantitative relationship in the *degree* of temporal coupling between storyteller/listener at specific cortical sites and the level of story comprehension by the listener. In other words, the extent to which a storyteller's brain and the listener's brain "ticked together" determined how successful the communication of the narrative would be (Stephens et al. 2010). Remember Polly Wiessner's observations in the Introduction about storytelling among hunter-gatherers, where she noted: "The language of stories tended to be rhythmic, ... with individuals repeating the last words of phrases or adding affirmations ... [frequently] they arrived on a similar emotional wavelength as moods were altered."

Another way to look at these findings is that they give a specific neural dimension to the belief that literature represents an intersubjective, mental event. Other researchers then used the same basic approach to identify brain areas reliably activated during story production *and* comprehension (Silbert et al. 2014). There were, of course, some areas specific to *either* one or the other. For example, motor areas for speech articulation were active only during telling a story; and a patch of parietal cortex, bilaterally, and the right anterior frontal gyrus were active only during comprehension. Yet a larger set of structures displayed activity during *both* story production and comprehension. In important controls, they showed that areas active during comprehension were not engaged by bouts of speech-like sound that lacked narrative content. Thus the neural activity recorded does indeed reflect narrative comprehension.

The global pattern of regions active during the experiment was itself important. While some active areas during production were localized to the left hemisphere, there was much bilateral activity. Finally, during both story production and reception, some of the classic language areas of the brain, such as Wernicke's and Broca's, displayed activity, in addition to other locations that will now be familiar—the precuneus and the medial prefrontal cortex—key sites that were seen to be active in the studies on Shakespeare described above.

As the research group was quick to point out, emerging data about these structures, part of the "DMN" (see Figure 6.3 below), suggest that "the ability of a listener to relate to a speaker and therefore understand the content of a complex real-world narrative seems to rest in these higher-level processing centers" (Silbert et al. 2014: E4693).

The neural implications are clear: we have indeed traveled "beyond Broca and Wernicke." The narrative implications of the notion of *brain-brain coupling*, provides for the first time a specific physiological grounding of what occurs when we share stories. At this point, it is not at all fanciful to say a good storyteller can entice listeners to "get on the same wavelength"— almost literally.

How Does Imagination Fit In?

We turn now to the "why question" lurking beneath the studies described. Why would people actually bother to form mental models of the sort proposed by cognitive theorists? A reasonable reply would be that we probably employ them not just for engaging with novels or films, but in everyday life, and that would explain why we came to have these ancient cognitive capabilities. If so, then we need to consider if such imaginative tools may assist and structure discourse in quotidian situations.

When looking at everyday conversation, strong evidence has been found that inference-making is a big component. As we all know, intended (and unintended) meanings even in the most informal exchange are often very different from what is literally said. ("What sort of time do you call this?") It should not be a complete surprise, then, that inference-making is associated with increased activity in many of the same brain areas mentioned above for processing dialogue in narratives (Jang et al. 2013).

Narrative cognition is a routine activity. Paul Harris, studying young children and their imaginative lives, asked: why are children driven to invent imaginary creatures and take part in "make-believe"? This question surely seems linked to issues around our propensity to construct and consume fictional narratives. It is known that the emergence of pretend play is closely intertwined with the development of language. In Harris' account, what connects these two phenomena is precisely the situation model. He states:

> There is now a wealth of evidence that when adults process a connected narrative, they construct—in their imagination—a mental model of the narrative situation being described. Moreover, as the narrative unfolds, they update that situation model so as to keep track of the main developments in the plot. (Harris 2000: 192)

Children seem to build similar representations to track changes in their make-believe worlds, and they can adopt the perspectives of play-partners to comprehend causal connections. Harris believes that the similarities in the way adults track narrative discourse and children track make-believe are more than a coincidence. In fact, he argues convincingly that when children reproduce a simple story they use situation models, and these have the dual function of allowing them to engage in pretend play and to process narrative discourse. Why otherwise would our biology allow us to cognitively abandon the "real world"? In the course of language development, children may increasingly need to construct a situation model not only when they listen to a story or fairy tale, but whenever they encounter connected discourse whose deictic center, or point of view, is displaced from the here and now.

For example, even at two or three years old, children share collaborative recollections with a parent or have conversations about planned events. A very young child will not have schemata in memory for things they have not witnessed, so has to imaginatively construct a situation model *de novo* and then update it as needed. A child who could only interpret discussions with others as relating to the present, and what can be witnessed directly, would be severely limited in their ability to comprehend language, or to learn about situations they themselves have not experienced.

So Harris concludes, "The cognitive capacity that underpins pretend play—the capacity to construct a situation model—is an endowment that enables children to understand and eventually produce connected discourse about non-current episodes." This hypothesis suggests that the evolutionary path to *Homo sapiens* depended upon not just the emergence of complex language, or the ability to create imaginative models, but a fusion of the two—allowing us "to exchange and accumulate thoughts about a host of situations, none actually witnessed but all imaginable: the distant past and future, as well as the magical and the impossible" (Harris 2000: 194–5).

The ideas of Paul Harris dovetail to a striking extent with the hypotheses of Michael Corballis and Daniel Dor, described in the previous chapter, where the practical utility of talking about past or possible events and characters may have been a driving force in language evolution.

A final point here about the relation of imaginative thought to everyday life: often the word "imaginative" is loosely associated with the fantastic, or other-worldly, but increasingly we can see that the narrative mind is not far removed from ordinary human cognition. Literary imagination is very much part of everyday life, as can be seen in most of our social interactions, and as recorded by Harris' work on childhood tales and childhood play.

But there are other compelling sources for such a belief. Molly Andrews, for example, has invested a career in listening to people's narratives, and has documented the value of narrative imagination in our work lives, education, politics, and even perception of aging. As she says, imagination "is not something that we dust off and put on for special occasions, a psychological tiara of sorts. Rather, it guides us from our waking hours to when we go to bed at night. It is with us always, sitting side by side with our reason and perception" (Andrews 2014: 11).

Scaffolding for Narrative Imagination?

Surprisingly, it was studies of the resting brain that first flagged the existence of unexpected regions involved in narrative appreciation and composition. The so-called DMN was revealed almost by serendipity, and its initial reception in the field was rather lukewarm (see Buckner 2012). The story of its chance discovery will help to explain its possible role in literary cognition.

Marcus Raichle, who has moved both PET and fMRI technology forward over several decades, has indicated that it was a chance observation from his research group that enabled functional imaging to be extended to a deeper range of cognitive questions. They observed with both imaging technologies that there was a decrease in activity at some brain locations during the performance of goal-directed tasks, compared with the resting state (Figure 6.3). As Raichle later pointed out, the resting state was not assumed to be a meaningful control for cognitive tasks and so these data were not usually included.

Despite this, "some did, indeed, include a resting state in their imaging studies and we were, unashamedly, among them!" And then, "at some point in our work, and I do not recall the motivation, I began to look at the resting state scans minus the task scans. What immediately caught my attention was the fact that regardless of the task under investigation, activity decreases almost always included the posterior cingulate and the adjacent precuneus" (Snyder and Raichle 2012: 904, Raichle 2015: 436). It turned out that such areas are relatively high in neural activity at rest, but become significantly less so as we are challenged to handle some types of mental work!

It is the group of brain areas shown in Figure 6.3A—especially the precuneus and posterior cingulate cortex (PCC) and the prefrontal cortex along the medial hemispheric surfaces (mPFC), and the inferior posterior parietal region on the lateral surface—that have become known as the

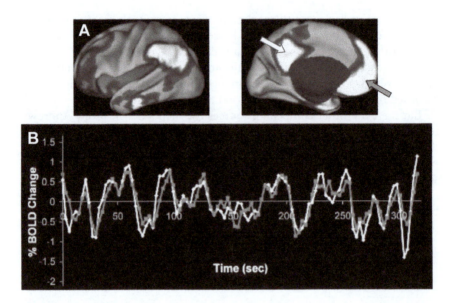

FIGURE 6.3 *Defining the default mode network. Panel A displays fMRI scans to show areas whose activity decreases during task performance. On the left is a view of the left cerebral hemisphere's lateral surface, a view of its medial surface is shown on the right. Bright areas are those showing the most substantial decrease during task performance. The large bright area on the lateral surface is the inferior, posterior parietal cortex (IPP). The bright areas on the medial surface include a posterior area, white arrow, centered on the precuneus and the posterior cingulate cortex (PCC) and an anterior area, gray arrow, centered on the medial prefrontal cortex (mPFC). Panel B shows fluctuations in the fMRI signal from these two areas over time with the white and gray traces corresponding to PCC and mPFC respectively. From: Raichle (2008) A brief history of human brain mapping.* Trends in Neuroscience 32:118–26, *Figure 4(a). Used with permission.*

DMN. They are not the only regions that show task-related decreases, but "serve as exemplars of all areas of the cerebral cortex that exhibit a systems level organization in the resting (default) state" (Raichle 2008: 124).

When we are awake but resting, or sleeping, our brains are continuously active at a high level. Then, when we undertake a mental task, the increase in energy consumption is only a matter of 5 percent or so. The unexpectedly large energy consumption during the "resting" state has been referred to as "the brain's dark energy." Moreover, not only do some regions hum along at a high level during rest, but they do so in a synchronous manner (Figure 6.3, panel B). These distinct brain regions "fire together" as a network, so it is a good guess that they share functional relationships.

What exactly is going on during rest to make the brain "run hot"? It seems that a substantial share of the brain metabolic energy consumed overall goes to support the intense work at synapses. However, at the level of the cortex most of those synapses (80–90 percent) are not positioned in pathways directly responsive to external sensation, but instead are part of entirely internal communication networks, like the DMN. This still somewhat mysterious entity is located in both hemispheres and generally toward the midline (the surfaces between the two hemispheres), unlike most "linguistic" areas we have considered so far (Raichle 2006).

Discovering the DMN led researchers to examine the resting state for other regions that cycle together, expecting there would be some which, unlike DMN, displayed low levels of activity during rest and increased during mental engagement. They found them. The work of a number of groups has resulted in the identification of a *dorsal attention network*, an *executive control network*, and several others (Chapter 3). These so-called task networks are often out of sync ("anticorrelated") with the DMN. So for example, the dorsal attention network increases activity during mental tasks, while the DMN decreases. All of this radically alters our conception of neuroanatomy, from mostly fixed "wiring," to a collection of dynamic interacting networks. Probing the default network, and others, has opened up an explicitly cognitive doorway to understanding "higher" brain functions and the human *connectome*: our brain's overall blueprint.

Intriguing suggestions have been made for the significance of the DMN. One possibility is that it is a core system for maintaining the brain at an appropriate "idle"—like a motor ready to react.[4] Other research has suggested that the DMN supports "internal mentation" such as mind-wandering and introspection, or that it may be related to a low level of attention—alert to novel stimuli but not directly focused on any one stimulus or task (Mantini and Vanduffel 2013, see also Corballis 2015).

Perhaps because of such varied conjectures, the idea of an intrinsic default network took a while to catch on, but has since become central to our understanding of cognition. Because the DMN now has a clear anatomical definition, it has been cropping up a lot in experimental studies, especially of language, linguistics, and narrative. Philip Davis' work on responses to Shakespearean functional shifts, and that of Hasson and colleagues on brains "ticking together" are two prime examples.

To take another, Regev et al. (2013) presented spoken and written stories to a group of people. They either heard a seven-minute narrative or read a transcript, while an fMRI scan was performed. As you might expect, early sensory areas responded accordingly: auditory when listening, visual when reading. Then there were some "higher order" cortical regions that responded to both (especially the superior temporal gyrus, inferior frontal gyrus, and the precuneus) probably reflecting basic linguistic processing. Strikingly, it is not just that these areas responded to both, but they did so with similar temporal dynamics.

Richard Wise and Rodrigo Braga have pointed out that several studies have seen evidence of DMN involvement in narrative processing, and one even found evidence for its involvement in narrative production. These associations of the DMN with literary response are intriguing because, as we will see, it also has an activity profile associated with memory and emotion. Furthermore, DMN seems to be a site where autobiographical memory interacts with social cues and mentalizing, and it's clearly where some aspects of discourse comprehension occur. In the chapters ahead, we will turn to memory, emotion, and sociality and their relationship to narratives more directly (Smallwood et al. 2013, Yang et al. 2013, Andrews-Hanna et al. 2014, Wise and Braga 2014). In doing so, we'll touch on what happens to such neuroscientific concepts when they turn up within novels with their own, highly specialized, and directive narrative means to look at what Jean-Dominique Bauby fancifully referred to as "the butterflies" inside our heads.

Notes

1   There is subcortical involvement of speech and language processing under some circumstances (e.g., Freiderici 2006, Simonyan et al. 2016).
2   Entries in our semantic lexicon do not have associated definitions like a physical dictionary but are thought of as being associated with lists of characteristics ("hero" might be connected with terms like brave, strong, etc.). These attributes help explain the way specific metaphors seem to make sense (see Hogan 2003b).
3   It is highly recommended you visit http://gallantlab.org/huth2016. There you can see the atlas and explore it with ease. It is worth the trip.
4   From a phylogenetic perspective, it would be odd if DMN had no specific functions. Well an equivalent network was found in monkeys and cats, and perhaps rats. That means DMN may be *evolutionarily conserved*, which suggests it supports important functions. As more has been learned about the resting state, it's become clear that DMN is not a homogeneous system, but a federation of several subnetworks: Andrews-Hanna et al. (2010), Raichle (2015).

PART IV

Converging Paths?

In the remaining four chapters, we turn to more detailed analysis of one type of narrative, literature, in relation to memory, imagination, and social and emotional cognition. This adds some useful examples of how the ideas covered so far may play out in fiction at the levels of theme, form, and character. How do writers incorporate and reflect them? We use works by Ian McEwan and E. L. Doctorow to illustrate some ways that neuroscience and the narrative arts can converge on fundamental explorations of the self.

7

Affective Cognition and Sociality

In the 1970s, astronomer Carl Sagan, writing for a popular audience, helped spread a myth about the neuroscience of affect (emotions, feelings, moods) that has penetrated broader culture. The idea was featured both in a book (*The Dragons of Eden*, 1977) and in a television series, where he promoted the idea that humans have a reptilian brainstem, an "R-complex," that triggers our emotions and echoes in our dreams where "dragons can be heard, hissing and rasping, and the dinosaurs thunder still" (157).

The notion originated with neuroscientist Paul MacLean, who speculated that mammals have three brains in one, a "triune brain." At its base was our brainstem which he believed was a "reptilian" vestige (hence, R-complex) mediating aggressive and survival behaviors. The second level he assigned to the so-called limbic system, an older term for structures at the very edge of or below the cerebral cortex. They were thought by some to be our primary emotional circuitry; MacLean described the limbic system as a veneer of primitive mammalian inhibition over our wild brainstem. The third and "top" level was the large and more recently evolved "neocortex" that he suggested oversaw the two lower levels. In short, McLean posited two levels of mammalian brain working to keep our relict reptilian emotions in check.

However, the idea of an unchanged reptilian module at the core of our brain is a serious misapplication of evolutionary thinking; there is no credible anatomical evidence for it. Mammalian brains have a mammalian brainstem. There is no principle through which it would be frozen in a reptilian state as the brain became adapted for mammalian lifestyles; and interactions between cortex and brainstem are not as simple as implied by the "triune brain." We have long known there are numerous bidirectional interactions between the brainstem and cortex and we formulate and control behavior through their subtle interactions (see Butler and Hodos 1996 and Striedter 2005 for critiques).

Despite the lack of evidence, if you google reptilian complex, it will return over 1-million hits, many from clinics offering help to control your "inner reptile," and marketing firms happy to train you to sell to that same fictional reptile lurking within customers. The example of the "triune brain" shows how a provisional idea about the brain can transform into a resilient meme—especially when it touches on emotion and is couched in simple narrative terms.

In this chapter, we travel deeper into the fascinating and contentious territory of affect and emotion. The treatment of these mental states, like many topics we have considered, reaches back to Darwin and even earlier, with roots in both Western and Eastern philosophy. It is also a vibrant area in neuroscience with fresh concepts challenging older theories. Antonio Damasio, for one, has worked to slay Sagan's dragon (*Descartes Error* 1994, and elsewhere), providing abundant clinical and experimental evidence that human rationality is produced by incorporation of emotion into decision-making.

Joseph LeDoux in reviewing the neuroscience of emotion (1996) suggested that the term "limbic system" should be abandoned—it is not a functionally unified system and some of its structures have little or nothing to do with emotion. Similarly, Lisa Feldman Barrett (2017) believes it's time to finally lay misguided notions of a triune brain to rest, and retire the term "limbic system" as it is presently understood.[1] In this chapter, we will uncover some more nuanced ideas on how affective cognition works, how it is intertwined with our sociality, and why this matters for a consideration of literature and storytelling.

Affect and Literature

One legacy from the New Critics is a concept called the *affective fallacy* that cautioned readers and critics to be wary of emotions and the "psychological effects" of fiction, lest they contaminate critical evaluation of a work (Wimsatt and Beardsley 1954). This admonishment has roots that can be traced to I. A. Richards' distinction between the symbolic and emotive uses of language, but he was actually very interested in psychological and emotional effects (see Chapter 1). After all, his model of literary reception postulated that words, once perceived, trickle through cognitive processes, where images are evoked and memories recalled, generating the colors of emotional tone.

The impact of the "affective fallacy" has diminished considerably, in part because of a devastating critical analysis by Stanley Fish (1970). However, the role of affective states in literary response is often acknowledged with a sense of unease even yet. As critic Jane Tompkins summed it up, "We have not been trained to focus on [a reader's] affective dimension. We are dumb in that we lack the skill to articulate in a publicly interesting and intelligible way the nature and structure and varieties of emotional response" (1977: 177). Within fiction, Ian McEwan's Henry Perowne (a neurosurgeon) in *Saturday*, a novel we look at in the next chapter, recoils from the immediacy of his emotions, taking flight into intellectualization to a comic extent. Henry endlessly examines the causes "proximal and distal" of his own emotional state. This strand extends also to E. L. Doctorow's character Andrew, considered in Chapter 10. Both fictional treatments are caricatures of an outdated fear and distrust of affect.[2]

Yet as these novelists know, one benefit of the recent "cognitive turn" in literary and narrative studies is that it seems to have reduced the level of suspicion about affect, and even fueled some fascination with it. For example, critic and theorist Norman Holland (2009) has said that for him emotion is the most intriguing and puzzling aspect of the literary response. In considering the phenomenon of "transport" into a storyworld, he asked how is it that the mental state of losing ourselves in a story helps us feel real emotions toward imaginary objects (i.e., fictional characters). He noted, though, that the scientific study of affect seems mired in a vast and often confusing terminology. This critique is spot-on, the terms "emotion," "mood," and "feelings" will require clear definitions ahead.

Affect and Psychology

Following up on Norman Holland's comment about the confusing terminology surrounding emotion, let's clarify terms. The noun *affect* is usually taken to mean the full range of our "valenced" experiences, that is, those with a discernably positive or negative tone. This general term then subsumes ordinary language words such as "emotion" and "mood."

No definitive taxonomy of affective states has yet emerged, but we will use terms largely consistent with the thinking of psychologist James Russell (2003), whose writing on the topic seems very carefully delineated. In brief, he says that "core affect" is an ongoing neurological state we consciously sense, having dimensions that can be represented as both valenced (along an axis: pleasure to displeasure) and arousal-related (along an orthogonal axis: calm to activated). Like Keith Oatley and Philip Johnson-Laird (2014), Russell considers core affect to be a psychological primitive, irreducible and universal. Universal here means that everyone has some behaviors characterized by these two axes, which can be used to map-out our affective states. Fear, for instance, would typically be located in a region very negative in valence and high in arousal.

Russell notes that core affect is called *mood* by some psychologists, but *feeling* by others. Core affect in his sense can be free-floating—that is, it need not have an object, or be "about" anything. So it does not display intentionality and Russell therefore considers it *mental, but not cognitive*: we will refer to such affective states, sometimes of extended duration, as "moods."

The only other primitive his account requires is our perception of this core affect—a process that is happening all the time. This framework gives the term "emotion" a distinct but related definition. Emotional events for Russell, like fear or anger, are *attributed affective states*: derivable from the two primitives, they occur when a change in core affect is perceived and linked to a cause (e.g., a person, place, or object). For this reason, unlike core affect, an emotion event is "about" something: it is related to an intentional object. Being intentional, he considers emotions then to be *both mental and cognitive* in nature. So affect is a broad mental category containing moods and emotions—which overlaps with cognition because of intentionality.

It would be natural to wonder: what might be an affective state that is *not* specifically attributed to some cause? Think of music. It certainly can evoke affects, but tying these to specific categories is hard. Indeed, it has recently been suggested that music communicates and evokes *affects (or moods)*, precisely because it evokes states with valence and intensity, but without an attributional quality (Cespedes-Guevara and Eorola 2018). This is an intriguing, if currently incomplete, explanation.

There is much disagreement in the field about how many emotions we have, whether or not they come in discrete categories, and whether they are inherited or constructed under the influence of language and culture. To come up with even provisional answers to these questions, science is necessary but not sufficient. We can only come to some initial conclusions below by considering a range of perspectives, including biology, psychology, language, and narrative diversity—all of which we believe are relevant.

Affect and Biology

Classic biological views, extending back at least to Darwin, tend to start analysis of affective states at the behavioral level. Emotions are most often defined as responses of body *and* brain consisting of arousal and preparation for action: stopping in your tracks, contorting your face, or turning to run if—using William James' famous example—you encounter a bear.[3] These reactive states are also associated with inner, subjective experiences, but crucially also with *expressions*—gestures or words that can communicate your mental/emotional state to others. They are allied with such basic behaviors as fighting, fleeing, or mating. Of course, the range of emotions as we speak about them in ordinary language is far broader than these stock examples.

The term "mood" is doubly useful because it is common in clinical neuroscience, where "disorders of mood" are the most common brain-based disorders: depression, bipolar disorder, and anxiety. These are long-lasting states compared to transient emotional responses, so moods align reasonably well with Russell's term "core affect." Russell and others also refer to core affective states as feelings. In recent neuroscience this term does not denote background states that precede emotional reactions, but rather refers "to the conscious experience of these somatic and cognitive changes ... *accounts our brain creates* to represent the physiological phenomena generated by the emotional state" (LeDoux and Damasio 2013: 1079, our italics). Two things stand out here. First, emotions involve both cognitive and somatic or bodily changes (more on that below) and second, feelings in this usage are associated with the process of establishing appropriate attribution for an emotional response. What makes *feelings* different is precisely their "aboutness."

Therefore, to arrive at three terms that seem distinguishable: we'll use **mood** for Russell's core affects (similar to its usage in psychiatry), **emotion** for reactions of brain and body triggered by definable objects or circumstances, and **feelings** for the conscious accounts we create subsequent to emotional reactions.

Russell suggests that when moods and emotional reactions are named individual and cultural differences enter the story. Theories of psychological construction (Barrett 2017) agree that cultural influences become evident at this point. Certainly, we do not always attribute our feelings to the correct cause. Without misattributed feelings, the world would lack most of its great dramas and all of its tragedies. There is no room for an emotionally perspicacious Othello or Oedipus.

Affect and Narratives

Patrick Colm Hogan, like Keith Oatley and others, has hypothesized that the majority of readers experience emotional situations in fiction by reference to their own emotional pasts. This suggests that specific autobiographical memories are easily cued by a story (see Chapter 9). Thus from a cognitive perspective, a storyteller or writer may guide a listener or reader on virtual journeys into their own emotional pasts, and construct salient emotional "landmarks" along the route.

But Hogan believes a small journey happens in those creating a story. "In following through the development of plot and character, the author too feels emotions, due to the priming of his/her own memories" (Hogan 2003a: 74). As we'll see, both McEwan and Doctorow draw on not only a deep reservoir of affective memory (both their own and ours) explicitly pointing to the inspirational recall of specific autobiographical incidents in the genesis of their work.

Hogan's reading of narratives across cultures and his resulting synthesis builds a case that our acquired understanding of emotion words—the semantics of emotion—is aligned with what he calls "prototypical situations," defining these in terms of inner narrative. Prototypes in his thinking may also be referred to as cognitive "schemata." In his formulation, "Our mental lexicons include prototypes for narratives and we judge narratives ... by reference to these prototypes" (Hogan 2003a: 87). He infers that our very grasp of emotional concepts derives from an inner "library" of prototypes—typical situations and responses he calls "mini-narratives." If we had never ourselves experienced fear of violence, or grief, or guilt, we would be deaf to characters as conjured by the likes of McEwan and Doctorow. It is noteworthy that Russell refers to emotions well known from folk psychology—fear, anger, happiness, etc.— as *prototypical emotions*.

Hogan's largest claim is that a small set of "core" emotional prototypes are human universals (Hogan 2010). As indicated in Chapter 1, he derived this position from extensive analysis of stories across many cultures and historical eras. Within this corpus, he noted recurring story schemata, including heroic epics, romantic comedies, power struggles, and stories of sacrifice. Of course this *question of universal processes* behind the particularity of individual affective states is one biology and psychology have long wrestled with, and is central to any assessment of narrative-inspired emotions. If there are human emotional universals, they might determine the reach of future theories of literary emotion.

Emotion Categories and Culture

The modern era of biocultural research on emotions is traceable to Charles Darwin, whose 1872 book, *The Expression of Emotions in Man and Animals*, began by summarizing knowledge of the muscles of the face, and their role in expression, based mostly on earlier studies of Sir Charles Bell and Friedrich G. J. Henle. He also reviewed experimental work by the neurologist Guillaume Duchenne that involved studies of patients who had lost sensation in the facial region. On this "blank canvas," Duchenne mimicked expressions by stimulating individual muscles electrically. The goal was to define how activation of specific sets of muscles could produce facial "gestures" communicating emotional state. This study reported that through combinatorial activation of the forty-two facial muscles, "authentic looking" emotional expressions could be reproduced.[4]

Duchenne's was an early reductionist approach to emotion, reasoning that if you can elicit facial gestures in a controlled way, there is hope of a systematic understanding of at least its expressive aspects. Facial expressions have ever since figured prominently in debates about categories of emotion and the claim that they might be human universals.

Darwin's legacy was a broad conceptual approach that still looms large in emotion studies. He examined communicative gestures in a range of animals to argue that affective states occur throughout the vertebrates and did not suddenly appear with the advent of *Homo sapiens*. Jaak Panksepp (1998) has been an advocate of this view, and so have others in the field. Panksepp (2011) further speculated that affective states in animals may have been a sort of precursor of consciousness, predating its "full" elaboration in humankind.

As for the detailed analysis of emotion, Darwin deduced that there were core or "basic" emotions: *grief, joy, anger, disgust, fear*—and depending upon how you read *The Expression,* perhaps *surprise.* To establish each of these, he used techniques that foreshadow modern experimental psychology. For instance, he used photographs of facial expressions, asking people to classify the nature of the emotional state pictured and found enough consistency to propose his basic categories, carefully including reports (mostly anecdotal) from more than a dozen non-European groups.

Studies of emotion have been cross-cultural since Darwin, and clearly an inclusive approach is necessary to avoid the biases of restricted focus on "unusual" groups. As pointed out by Ara Norenzayan and Steven Heine (2005), most data for psychology experiments come from the US, the UK and Canada, and also tend to come from undergraduates at university: a sample that can be characterized as largely White, Educated, Industrialized, Rich, and Democratic, or more simply, WEIRD (Henrich et al., 2010). Unsurprisingly, this group has proven unrepresentative of humankind.

The range of cognitive/behavioral functions whose descriptions depend upon the demographic profile of the group being studied is large, including: analytic versus holistic reasoning, the use of numerical concepts, ethical decision-making, concepts of self, and even aspects of basic perception. Such cultural variability requires that any applications of empirical research in literary studies must be mindful of the datasets used to avoid premature generalization. Henrich and colleagues have summarized how resolution of the issue of culture versus biology in emotion ends up feeling like the resolution of the old nature versus nurture debate. "Recognizing the full extent of human diversity does not mean giving up on the quest to understand human nature. To the contrary, this recognition illuminates a journey into human nature that is more exciting, more complex, and ultimately more consequential than has previously been suspected" (Henrich et al. 2010: 29).

Mindful of the power of language, we will refer to Darwin's six categories as "typical" emotions. They match up reasonably well with Russell's prototypical emotional responses, and the emotion terms associated with Hogan's emotional prototypes from world narratives. We are not implying that these categories are psychological primitives or universal—as tends to be the case when using the term "basic emotions" (e.g., Ortony and Turner 1990, Panksepp 1998).

The Face of Emotion

In a classic update on Darwin, Paul Ekman (Ekman et al. 1987) had subjects from ten different cultures rate photographs of the typical emotions: Darwin's six plus a seventh, "contempt." They reported strong agreement across cultures in the "reading" of the emotion category displayed. Nevertheless, opinions about the intensity of emotion were less consistent, and a post-hoc analysis indicated that there was a discrepancy between Western and Eastern cultures. This was taken to mean that these emotions may be human "universals," but that there are some culturally specific influences in their interpretation, a conclusion congruent with work by psychologist Carroll Izard (Izard 1994).

While the claim for emotional universals has been around a long time, some in the field remain unconvinced. There are at least two good reasons for this. First, consideration of the methods used in the field—showing photos of faces and asking respondents to assign words the experimenter supplies, or providing a story-like scenario, then choosing which of two faces match—tends to force a response without considering the concomitant effects of words and stories on perception. In fact, providing verbal prompts tends to create agreement of the sort Ekman reported; avoiding prompts does not (Barrett 2017). Second, some studies testing for universal emotional perception with rival methods have produced divergent results. A recent example comes from a report by Jack and colleagues. Using a digital library of facial expressions, they collected data showing that Western subjects recognized emotions that could be fitted into six (putatively "universal") categories, but individuals from Eastern cultures did not fit the expressions into the same categories (Jack et al. 2012, Jack et al. 2013, Sauter and Eisner 2013).

One recent study suggests how we might begin to make sense of these results. Masuda and colleagues found that when studying a face, Japanese people were influenced in their rating of emotional intensity by expressions on the faces of *other* individuals seen in test pictures. Eye movement tracking revealed that the Japanese participants spent more time scanning surrounding individuals than the Westerners—that is, there was an attentional difference. This means that tests of emotion perception with some Asian subjects that use one face in isolation may be artificial for them. Remember that a cultural difference in perceptions of emotional intensity was observed in Ekman's study, cited above, and all of this may indicate a more holistic form of perception by Japanese people than is typical of Europeans (Masuda et al. 2008, Masuda et al. 2012).

In brain-based work, subjects viewed pictures of faces as researchers adjusted the images so that they morphed from one expression to another, meanwhile simultaneous fMRI scans were made to see if brain activity in the fusiform face area (FFA) of the temporal lobe (see Chapter 4) would change correspondingly. When facial expressions were made to smoothly shift from one emotion to its apparent opposite, say sad to happy, there was a point at which perception suddenly "clicked" from the first to the second category. This happens in many cases of perceptual judgment and is known as "categorical perception."

The FFA showed a distinct shift in fMRI signal just at the point where an observer reported that her perception jumped from one emotion to the other, consistent with the idea that facial perception is "chunked" into distinct emotional categories. Intriguingly in the same experiment, several brain sites related to emotion displayed a different, more continuous fMRI pattern— one that did not shift at a category "boundary" but simply correlated with perceived emotional intensity (Matsuda et al. 2013).[5]

These results are consistent with the notion that we perceive emotional expressions in a categorical way, but they also suggest that we use a distributed set of brain locations to do so, and that several events are occurring simultaneously. Some neural loci are processing facial cues that probably reflect valence, while others are processing cues for intensity judgments. The net effect is a subjectively reported sense of distinctive categories. These events in their totality seem best characterized as "affective cognition."

We have taken the time to cover the perception of emotion from facial cues because it provides abundant data on the important questions of emotional categories and whether culture influences affective cognition. But bodies also contain information related to emotional state, viewers do not always rely on facial information alone; indeed, under real-life conditions emotion may be detected solely from body dynamics (Heberlein et al. 2004) perhaps even more successfully than from the face (Aviezer et al. 2015, Abramson et al. 2017). Of the negative emotions, anger seems especially detectable from body cues, perhaps reflecting the biological imperative to detect this state from a safe distance (Martinez et al. 2016).

What about other sources of emotional cues? As mentioned in Chapter 4, voice tone is represented early, and selectively, at the cortical level (e.g., Belin et al. 2000, Moerel et al. 2012). This makes sense. Voice encodes aspects of emotion that are communicated concurrently with facial and body expressions (e.g., Magnee et al. 2007). The impact of the voice either can be due to the occurrence of certain frequencies characteristic of non-speech sounds (Pisanski et al. 2016, Raine et al. 2018) or can depend upon the prosody of speech (e.g., Witteman et al. 2012, Iredale et al. 2013). In short, emotion recognition is multisensory and uses many contextual channels related to face and body dynamics, speech, and non-speech sounds, providing many affordances for cultural impacts on our neural processing.

Studies of facial, body, and voice cues all show evidence that people can place emotions in categories, and not only that there is an ability to recognize emotional expressions across cultures, but also that there is cultural variation. Can neuroscience add anything to help clarify the situation?

Neural Networks in Emotion and Feeling

Early research to uncover the parts of the central nervous system (CNS) involved in emotion indicated that behaviors associated with emotion depended less on the cerebral cortex and more on subcortical sites—for example, the hypothalamus. This led to the general idea that subcortical regions drive somatic components of emotion reaction based on activation of autonomic nervous system (ANS) pathways that control breathing, heart rate, and sweating. What cortical processing then adds is our conscious awareness of feeling states and also a route of influence for linguistic labels.

Figure 7.1 summarizes some key components, but by no means all, of the CNS areas controlling emotional reactions—note the hypothalamus at the base of the diagram. The hypothalamus and the nearby "central gray" region of the brainstem drive observable bodily concomitants of typical emotions: freezing (or in some cases a startle response) autonomic effects on blood circulation, respiration, and hormonal changes. These effects prepare the body for action—getting blood to the muscles, increasing hormones that will enhance metabolism, etc. And they can extend beyond the "emotional moment," as we will see in the aftermath of Henry Perowne's car-crash in the novel *Saturday*.

A key controller of the hypothalamus in cases of emotional arousal, the amygdala, can be thought of as a second level of emotional control tucked beneath the temporal lobe. Emotional stimuli activate the amygdala directly just as they activate cortical sensory zones. The cortex, in turn, also activates the amygdala. The direct, non-cortical route is one way the

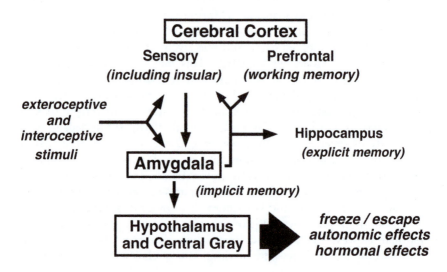

FIGURE 7.1 *Affect-related areas occur at several levels of the CNS and have points of connection to memory. Only key structures are shown at three different CNS levels (boxes). Sensory input can access the amygdala directly or after relay through the cortex. Basic motor outputs (bottom right) are driven by the hypothalamus and the central gray. Implicit emotional memories are stored in the amygdala and probably elsewhere. Amygdala output pathways enable working memory storage in prefrontal cortex and explicit memory storage in hippocampus. The hippocampus is not believed to have a direct role in emotion.*

amygdala participates in unconscious emotional processing and learning. It is faster than the pathway that passes first through the cortex—beating this indirect route by perhaps 10–15 milliseconds in the case of acoustic input. As neuroethologist Kenneth Roeder observed when discussing the widespread evasive behaviors of animals, "a millisecond or so within the nervous system must often mark the difference between the quick and the dead" (1959: 290). Such sharp reactions are obviously vital in dealing with threats, things that "go bump in the night." It is now established from direct electrophysiological recordings, even at the single neuron level, that the amygdala responds to fearful faces, and, remarkably, will do so even when they are presented so briefly that they are not consciously perceived (Morris et al. 1998, Whalen et al. 1998, Oya et al. 2002).

Given this unconscious sensory access to the amygdala, it has a large role in implicit learning (of which we are unaware). Learned associations can also be made regarding facial expressions, body postures, and vocal sounds. Such conditioning is an important role of the amygdala, and can be quite subtle in humans: we can even learn fear from language, or from observing others being conditioned (LeDoux 1996, Wilensky et al. 2006, Olsson and Phelps 2007). However, it is vital to state that this relatively compact brain structure is very clearly not "dedicated" to fear.

At the cellular level, it is differentiated into distinct regions and contains a number of unique neuronal populations with connections both internally and outside itself. Notably, it is activated when decisions are being made about the reward value of a stimulus. Emotions of positive valence have also been associated with it, becoming activated when subjects view pictures associated with food, sex, or money (e.g., Adolphs 2013, Berridge and Kringelbach 2013, Kim et al. 2017). Finally, its relationship to memory (emphasized in the figure above) is not just about implicit forms such as conditioning, but also about explicit forms, like declarative memory (e.g., de Voogd et al. 2016 and Chapter 9).

Sitting over the two subcortical levels shown in Figure 7.1, a number of regions of the cerebral cortex participate in emotion, including the anterior cingulate cortex (ACC), insular cortex, and ventromedial prefrontal cortex (vmPFC) (Figure 7.2).

Ventromedial prefrontal cortex is particularly well known because when it is damaged there are dramatic changes in emotions and personality. The celebrated story of Phineas Gage who sustained a substantial lesion in this area from a railway construction accident (1848) is particularly compelling. A fine retelling is available in Antonio Damasio's book *Descartes' Error*, and the incident was brought up-to-date by Hanna Damasio and colleagues when they remapped the likely site of the lesion (1994). This is reminiscent of the modern remapping of Leborgne's brain that was studied in the 1860s by Paul Broca (described in Chapter 4).

Analysis of Gage's skull indicated that the likely areas of damage were all in the frontal lobes, and especially extensive within the vmPFC (prefrontal

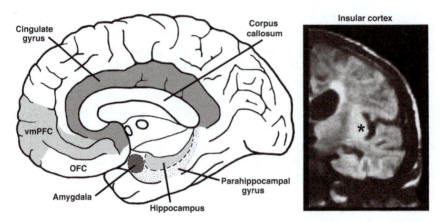

FIGURE 7.2 *Some key cortical areas related to affective cognition.* **Left:** *View of right hemisphere along the midline, rostral (anterior) is to the left. The amygdala and hippocampus have dashed outlines to indicate they are just below the medial surface. OFC = orbitofrontal cortex, vmPFC = ventromedial Prefrontal Cortex.* **Right:** *Structural MRI cross-section through about the middle of the brain shown, midline is left, lateral surface to the right. Insular cortex is marked with an asterisk.*

cortex) on both sides (Figure 7.2). Neuropsychological data has shown vmPFC damage compromises how emotions are processed and creates difficulties with decision-making.

In Gage's case, there also appeared to be some damage to the ACC. Such damage would also be consistent with effects on emotion. A fairly long-standing view of ACC has been that it is involved in "cognitive control" and this really means monitoring behavioral performance and signaling when predicted outcomes do not occur (e.g., Carter et al. 1998). Recent neurophysiology and modeling studies have suggested it may be involved in signaling valence and a sense of surprise based on performance discrepancies (e.g., Apps et al. 2016, Alexander and Brown 2019). At the same time, there is a body of experimental work that suggests ACC is related to social behavior, tracking social engagement with others and making social decisions (e.g., Apps et al. 2016, Lockwood and Wittmann 2018).

At this point, it should be obvious that a general inference is staring right at us. All of the main emotion-related regions are implicated in *multiple* behavioral functions, some highly related to emotions, some less so.

However, we should not take this to mean that affective states and emotional perceptions are represented everywhere, but rather that there is no one-on-one relationship between affective states and brain areas. There *is* no fear center, no happiness center, in our brain (see LeDoux 2017). The neural regions we have mentioned are part of several large-scale intrinsic networks, such as the salience and default mode network (DMN).

The evidence here is strong. Emotions and their percepts are not localized to specific nerve cells or brain nuclei; they are *emergent properties* of distributed affective-cognitive networks. Perhaps this is most easily seen using a computational analogy: many brain regions seem to be specialized in running a few particular algorithms, and a state like fear draws upon multiple algorithms, both reactive and predictive, to accomplish a goal. Behavioral attributes of "the person" cannot be reflected at the level of one brain region or individual neurons. What neuroscience has to do, and is very busy doing, is to figure out how neuronal activity waxes and wanes across many locations, drifting in and out of coordinated responses that represent the agony and the ecstasy of our affective lives (see Barrett and Satpute 2013, Kragel et al. 2016).

It is important to be clear that when speaking of emotions, we are often talking about two concurrent events: our subjective states and our observable behavioral/physiological reactions. Keeping these related phenomena distinct is important operationally to science and intimately dependent on language. However, there is still a logical organization to the way emotions are represented or enacted in our brain: "Objectively measurable behavioral and physiological responses elicited by emotional stimuli [are] controlled non-consciously by subcortical circuits, such as those involving the amygdala, while the conscious emotional experience [is] the result of cortical (mostly prefrontal) circuits that contribute to working memory and related higher cognitive functions" (LeDoux 2017: 303). This point is crucial. When the amygdala has been destroyed by disease, physiological defense responses may be compromised (e.g., Phelps 2006), but such individuals can still have experiences of fear and panic (Anderson and Phelps 2002, Feinstein 2013). Clearly, moving toward a better understanding of affect requires great care around the terms used, and a heightened awareness of when we are talking about overt behavior as against subjective experience.

Life on "the Island"

There is a final point to be made about the relationship of brain to behavior and subjective experiences and for that we'll briefly consider the insular cortex (Figure 7.2, right). The insula (Latin, island) and to some extent the cingulate cortex have been shown to have interesting associations with core affect (mood) and perhaps feelings. The insula receives substantial input from the viscera by way of the brainstem: information on the heart, lungs, and other vital organs is relayed to this cortical area (which you can't see from the surface).

It is perhaps aptly described as the primary sensory cortex for our *interoceptive* systems: our monitoring of internal organs, as opposed to exteroceptive senses, vision, hearing, touch, etc. Fascinating studies have revealed that neural activity in the right anterior insula, in particular, is high in those very sensitive to their own heartbeat (one way to assess interoceptive awareness). Indeed the very size of the insula and its physiological activation positively correlate with the subjective awareness of bodily feelings: people with a larger, more active insula are more aware of "gut" feelings than others (Critchley et al. 2004).

These results show that neural systems for affect can be lateralized with a right hemisphere advantage, the opposite of what we saw in Chapter 4 for language systems. This pattern of lateral asymmetry is also true for the amygdala and prefrontal cortex—the right side typically displays more involvement in feeling states than the left. The facts about the insula are generally consistent with the idea that affect is embodied—in the sense that your mental state is influenced by your viscera, moment by moment (Craig 2004, Garfinkel and Critchley 2016).

A biological understanding of these observations is highly relevant to affect and emotions in general, but especially to a consideration of our relationships to language and stories. Variability within populations as mentioned in Chapter 3 applies here: individual differences in emotive behaviors and experiences were to be expected. What was not expected was how such differences could be related to brain structures—the very idea that the size of a brain region would vary from person to person reflecting our idiosyncratic emotional traits was surprising.

There is a term in neuroscience we've used earlier that is helpful here: the phrase "brain space," to denote differentials and flexibility in the size of neural structures. Some of the examples are extraordinary; for instance, male songbirds dramatically increase the size of their song control nucleus in the forebrain every spring as they reactivate their song repertoire (Nottebohm 1981). As seen in Chapter 2, recent work found a relationship between the mentalizing skills of individuals and the volume of a region of prefrontal cortex. There are yet other instances (Chapter 9), but the point is that "brain space" is a plastic property: the thickness of cortical regions, and the density of neurons, reflects individual behavioral and cultural practices. Subtle neuroplasticity at the cellular level has long been known, but finding these flexibilities at the level of gross brain structures was most unexpected.

Language, Narrative, and Affective Cognition

Returning to literature, let's look at the evidence for what readers and critics have often assumed: that our responses to narratives include affective states, and in turn our processing of stories is influenced by our moods or emotions. Philosophers, aestheticians, psychologists, and critics have speculated long and hard on the details without reaching consensus. A full description and complete theory are not yet available. In recent years, however, some intriguing findings have begun to emerge.

Consider a study reported by David Miall (1989). He had a group of students read the introductory section of the short story by Virginia Woolf *Together and Apart* (1944). He was interested in testing theories, which posit that as we read we formulate "situation models" of key narrative elements (e.g., causes and goals of characters actions, see Chapter 6). Woolf's story describes two people introduced at a party, but the language of the first few sentences is ambiguous, and develops a metaphor whose import would not be entirely clear on first reading. What he found was that on initial reading detailed reactions to each line of the text could be rated for perceived importance to the story and affective intensity.

However, when subjects had read the entire story, their assessment of the situation model (concerning what was really happening in the text) changed and, crucially, it seemed to do so under the influence of passages with high affective intensity. His point was that if one applies classic, purely cognitive models to reading, you would be missing a channel for understanding real literary texts—in all their complexity, with embedded ambiguities, metaphorical associations, focalization, emphases from foregrounding, etc.—unless you factor in affect. Indeed he was making the case that reading a text is not just constructing a mental model of the words and sentences, but that full comprehension is really a process of creating situation models, storing them as schemata and updating them over time, and that affect is "a key agent" in this imaginative, flexible, and sometimes retrospective process (56).

Since Miall's study was published, there have been many others attempting to tease apart the components of affective-emotional responses and how they influence our wider cognition. For example, studies of the main brain structures mentioned above have indicated that beyond any associations with general cognitive tasks, all of these areas also are activated when we read (e.g., Herbert 2009, Engström et al. 2015, Lai et al. 2015).

Moving from the importance of affect in our narrative lives, we turn back to language itself and words. Recent work at this fundamental level has added some insight to one of the thorniest questions in the field: are human emotions universal, or is cultural variation really the starting point of all we should consider?

Well, a multidisciplinary team spanning comparative linguistics and neuroscience has completed a large review of a database of words and "colexifications" across a global sample of over 2,400 languages representing twenty language families, with highly revealing analyses (Jackson et al. 2019). Colexifications are cases where multiple concepts within a linguistic domain are co-expressed by the same word form within and across groups. An example of the thinking here might be this: two groups are likely to colexify concepts such as "water" and "ocean," but would they also colexify "anger" and "grief"? Cleverly, the study used the phenomenon of colexification to come up with a metric of semantic distance between words conveying emotion concepts in—say Indo-European languages compared to Austroasiatic languages—creating, from such comparisons, statistical "maps" of cultural semantic relatedness.

If cultural and psychological construction is really the only thing that should matter for the way we speak of emotions, then colexification across groups should be explained by the degree of geographical separation, because borrowing of words is known to occur between groups. What they found was that variation in emotion semantics was indeed correlated with the proximity of language groups. This strongly suggests that cultural variation is real and an important contributor to emotion concepts.

However, that is not all. They also sought to determine if any psychophysiological parameters could be used to quantify and predict the way emotion concepts were clustered across the colexification networks they found. Two dimensions, valence and arousal, were the only two of six factors tested that could do so. These two represent biologically based (*and* brain-based) factors that appear to organize our affective behaviors as described earlier.

In short, this study indicates that how human groups across the globe ascribe words and meanings to emotions shows variations *and* regularities in a way that suggests influence from both cultural and biological mechanisms. This conclusion strongly parallels that from the old nature-nurture chestnut: it is not one or the other, but a blending of culture with biology.

Social Emotions

Human lifespan is characterized by extended parental care that, in turn, allows for extended social learning. In the past fifty years or so, there has been a scientific shift from focus on extrinsic/ecological factors as shapers of human evolution to intrinsic and cognitive ones. So in addition to noticing our extended infancy, clever hands, and use of fire, there is now more focus on our clever minds and our unique form of sociality.

Living in groups requires specific mental skills, and research suggests that we quickly learn to understand and predict the behavior of other people by inferring mental states such as beliefs and intentions: hence our regular and compulsive engagement in "mindreading" (Premack and Woodruff 1978). Taking seriously the social implications of mindreading, it has even been proposed that we came to occupy an ecological niche defined less by climate and food availability, and more by neural and cognitive advances, enabling us to collectively and globally exploit resources in novel ways: we seem to occupy a "cognitive niche" (Tooby and DeVore 1987).

Refinements of such ideas have led to emphasis on the size of our social groups and the importance of language—all consequences of our long period of social learning. The "pooled knowledge" within a tribe gives individuals vast amounts of information both about the environment and about other group members. So, for example, during our hunter-gatherer stage when we had to compete with large predators, humans gained advantage through efficient social coordination, speed, and group intelligence (Pinker 2010, Whiten and Erdal 2012).

Thus it seems clear that our large brains enable this high degree of sociality together with our complex language—something that had been dubbed the "social brain hypothesis" (Dunbar 2003). Empirical evidence, backed up by studies of other primates, underscores how fundamental sociality is to our nature. For example, savannah baboons in Kenya have been followed for many years and a substantial body of data shows that females with more numerous, and stronger, social relationships with other females have superior survival of offspring compared to the average (Silk et al. 2003, Silk et al. 2009).

The basis for linkage between strong sociality and enhanced reproductive success is known from work on various primates. When two members of a group spend considerable time in a typical primate social behavior—grooming—it leads to an alliance. Having a network of social allies leads to assistance in childcare and enhanced infant survival for females, and facilitates hunting or access to mates for males (reviewed by Silk 2007).

Considering the dramatic social dynamics of our primate relatives, it is notable that their cooperative behavior is relatively limited in scope compared to us. For example, most altruistic responses are restricted to kin and directly reciprocating partners. By comparison, human social behavior is much more inclusive and varied (Pinker 2010, Silk and House 2011). This gives us some idea of the *why* of human sociality, but what of the *how*? What cognitive specializations account for intense sociality? Any answer needs to account for how *our sociality is entwined with emotions* and again we ask, what can neuroscience add?

Without doubt, it is relevant that a host of studies and meta-analyses have documented substantial overlap of the brain regions that are known to be important for mentalizing, with those engaged during emotion and with regions used in responding to stories (e.g., Mar 2011, Altmann et al. 2012).

To be sure, it is not that these areas are identical; for example, within the mentalizing network, there are some locations that are preferentially activated during "mindreading" people's emotions, with other areas most active when making inferences about their beliefs or thoughts (e.g., Molenberghs et al. 2016). Areas of the "mentalizing network" that are also active during processing of emotional stimuli include medial PFC, ACC, several temporal lobe components (middle temporal gyrus, superior temporal sulcus [STS], the temporal poles), the precuneus, and the temporo-parietal junction.

You may recognize several of these from our discussions of narrative comprehension, as well as the DMN. One particularly clear example should suffice to show their close interconnections. Deen et al. (2015) performed a detailed fMRI analysis of the STS and showed that this one region—on both sides of the brain—contains an array of overlapping zones each specialized for handling different but related functions: mentalizing, identifying faces and biological motion, recognizing voices, and processing the language in stories (see also summary in Beauchamp 2015). Clearly, the future of neuroscience will require probing the nature of our mental reactions to stories in terms of ever more specific affective cognition loci within our large-scale brain networks, a deeper and more particularized form of mapping.

Affective Cognition and the Experience of Narratives

As mentioned in the Introduction, emotion has often been characterized as quite distinct from cognition. However, that division has lost credibility (e.g., Phelps 2006) due, in part, to an appreciation of the clear intertwining of affect and cognition on a wide array of behaviors in our personal and social lives. In the study of stories, Patrick Colm Hogan, amongst others, has brought "affective narratology" into focus, and a stance on affect is proving essential to such areas as discourse processing, stylistics, and literary aesthetics (Hogan 2011b, Stockwell 2009).

In fact, the reading of fiction provides a powerful example of the interplay of affect with the more traditional aspects of cognition—and one that all of us are familiar with. "Literature not only depicts and provokes emotional experience; it contributes to the formation and operation of our emotion systems ... [it] is a valuable source of information about that contribution" (Hogan 2011a: 288). Hogan then surveys the many critical views of *Madame Bovary*, a novel which many remember for its depiction of how reading stories apparently contributes to Emma Bovary's downfall, arguing that the emotional content of her reading gave her imaginative access to a view of her marriage as desperately loveless.

Contrast this with the study of Jane Austen's novels and their impact on empathic understanding as described by William Deresiewicz; in his case, he found engagement with her novels to be emotionally constructive in the development of more positive relationships with others (Deresiewicz 2012). This is a point that Henry, the neurosurgeon in *Saturday*, arch-materialist, fails to grasp, dutifully reading *Anna Karenina* and *Madame Bovary* at his daughter's behest, retaining only a sense that "adultery is understandable but wrong, that nineteenth-century women had a hard time of it." He is unable to give the empathic investment required of him, remaining unmoved, since "a man who attempts to ease the miseries of failing minds by repairing brains is bound to respect the material world" (66).

You Are What You Read?

What seems very clear is that not only can we usually recognize typical emotions such as fear, sadness, happiness, and anger, but social emotions are widespread, and represent a large part of what we chat about amongst ourselves and the tales we tell. The words for essentially social emotions in language are extensive: embarrassment, guilt, shame, pride, empathy, etc. Centuries ago, Dr. Johnson, in his preface to Shakespeare, described the way literary "imitations produce pain or pleasure, not because they are mistaken for realities, but because they bring realities to mind" (in James Wood 2008: 239). In this final section we consider how narratives, literature, might foster our fellow-feeling, empathy, by bringing it to mind.

We ask first, have narrative effects on empathy been demonstrated by psychologists? In 2006, Raymond Mar, Keith Oatley, and colleagues published a research paper with the catchy title: *Bookworms versus Nerds: Exposure to Fiction versus Non-fiction, Divergent Associations with Social Ability, and the Simulation of Fictional Social Worlds*. This turns out to be a wonderful exercise in myth-busting and brings together several lines of theory to test the idea that reading fiction is a good way to engage with and enhance social skills (Mar et al. 2006).

As the authors point out, we tend to associate a "bookworm" with the stereotype of a socially awkward nebbish. However, based on ideas of how narratives engage us cognitively, they predicted regular readers of fiction would be the opposite: adept at handling, or at least understanding, social situations. This conjecture was predicated on the research of Jerome Bruner, Richard Gerrig, and Keith Oatley suggesting that literary narratives transport us into a realm of thought that mimics real social interactions: engaging in emotional simulations. Contrary to stereotype, Mar and Oatley proposed that it is heavy readers of non-fiction who are more likely to be "nerds," because non-fiction is expository and, unlike fiction, should not exercise emotional cognition skills.

To test their prediction they used an author-recognition test designed to provide an objective measure of familiarity with the two genres—fiction and non-fiction. Then they administered several psychological tests including a questionnaire to test empathy, and the "Reading the Mind in the Eyes" test (RMET), which challenges participants to "read" the mental state of people from photographic views restricted to the eyes. Significantly, those with reduced social perception (such as autistic individuals) do not perform well on this test.

What did they find? Well, degree of exposure to fiction was positively correlated with scores on the RMET test, as were measures for the amount of absorption in stories, social acumen, and the ability to take another's perspectives. However, exposure to non-fiction was not correlated with any of these measures. The findings are an elegant illustration of the possibility:

Bookworms, by reading a great deal of narrative fiction, may buffer themselves from the effects of reduced direct interpersonal contact by simulating the social experiences depicted in stories. Nerds, in contrast, by consuming predominantly non-narrative non-fiction, may fail to simulate such experiences and may accrue a deficit in social skills as a result of removing themselves from the actual social world. (Mar et al. 2006: 705)

Naturally, one test like this does not establish a principle. No experiment is perfect and this one is not, by its design, able to determine causal direction. So, habits of reading fiction correlate with good social skills, but one can't be sure whether the habit of reading fiction enhances or rather results from the skills: perhaps people with good social skills simply consume a lot of fiction. But the results certainly did provide impetus for future study.

Additional work has continued to demonstrate relationships between reading fiction and empathy. First, P. Matthijs Bal and Martijn Veltkamp made their measurements of empathy across the experiment, making them first before reading pieces of literature, and then up to one week afterward. They divided their group in two: one set read fiction, the other non-fiction. Intriguingly, they found engaging with fiction increased empathy scores after one week, but only if the members of the group had indicated they experienced "transport" (being moved by the story) while reading. Lack of transport actually led to lower empathy scores after one week (Bal and Veltkamp 2013).

Then David Kidd and Emanuele Castano assessed possible differences between literary fiction and popular fiction (admittedly difficult categories to define; they used award-winning and canonical works for literary fiction, and best-sellers for popular fiction). They reported that one session of reading literary fiction led to better performance on tests for mindreading compared to popular fiction, non-fiction, or reading nothing (Kidd and Castano 2013). It seemed quite extraordinary that just one session of reading fiction would improve a person's ability to detect mental state from observing someone's eyes (the RMET test). So, this experiment was, quite reasonably, subject to replication attempts. These have not so far upheld their main finding. However, all of the studies so far that also tested for *lifetime* exposure to fiction (through the author recognition test) reported an association with empathy.

One of the papers reporting an inability to replicate Kidd and Castano, but that did report a significant relationship between lifetime exposure to fiction and mentalizing was Panero et al. (2016). Ellen Winner was one of the authors of that paper and she has provided a cogent analysis of how we can approach the tricky and unresolved issue of empathy and fiction-reading. She calls the assertion that reading novels makes you better at reading emotions merely from viewing someone's eyes—the RMET test—a "far transfer" claim (Winner 2019). By this, she means it is a stretch to expect a work like *Madame Bovary* to improve the mentalizing task of detecting fine features of a person's facial expression around the eyes when the novel does not render emotion through these means. But perhaps, she suggests, there might be what she calls "near transfer," where say a narrative with focus on compassion would make readers behave more compassionately.

When such connections were sought, they did indeed emerge. For instance, reading stories that focus on a character behaving pro-socially did lead to a significant tendency for readers to provide help to someone in need afterward if—get ready for it—they reported being transported by the story (Johnson 2012). Other research structured in this way, looking for near transfer effects of reading, noted them in a case where people read about a woman facing and standing up to racial hatred, and another where children and older students reading Harry Potter exhibited improved attitudes toward stigmatized groups if they identified with Harry, rather than the negative character Voldemort (Vezzali et al. 2015). Ellen Winner places such findings in context with the distinction between what she calls "compassionate empathy" (really sympathy) from more generic formulations of empathy in a purely cognitive framework.

Studies in this field have also, inevitably, included brain imaging. This makes sense, since we already know that reading often activates the DMN, as does engaging in acts of social cognition, which would involve mentalizing (above and Chapters 2 and 6).

A recent fMRI study has shown that frequent readers of fiction displayed not only more social cognition, but also stronger activation within the dorsomedial prefrontal cortex region of the default network. Interestingly, a medial temporal lobe area was also strongly activated by the most vivid passages, but not tied to social content. The authors of this study concluded it "demonstrates that fiction reading recruits the DMN because it elicits at least two distinct types of simulation: the simulation of vivid physical scenes and the simulation of people and minds." These results open up the possibility that fiction "exercises" one's mentalizing by way of passages simulating social tasks (Tamir et al. 2016: 7).

There is much more to be figured out about the complex relations of emotion to various dimensions of our "mindreading" skills and to memory. Yet as progress continues, we should also remind ourselves that "understanding how we see or speak does not debase what is seen or spoken, what is printed or woven into a theatrical line. Understanding the biological mechanisms behind emotions and feelings is perfectly compatible with a romantic view of their value to human beings" (Damasio 2003: 164). We will now turn to the role of emotion not only in relation to others, but with regard to self-perception as played out in two novels showing a high degree of scientific awareness.

Notes

1   The term "limbic system" is still in wide use among clinicians. It refers to a set of defined brain structures, but they do not form an integrated system.
2   We note that the field of affective studies, in its emergence, has presented a theoretical counterpoint to cognitive studies. See, for instance, Gregg and Seigworth (2010).

3   Although encountering a bear may be uncommon in London, Dublin, or New York, it is not unusual in the US state of Montana. Here people know that whether it is wise to freeze or run from a bear depends upon the type of bear encountered and slowly backing away is often the best response.

4   For contributions to our historical understanding of emotion prior to the twentieth century, and the role writers played in developing emotion concepts, see Richardson (2001, 2010b).

5   There are several face-selective areas in our brain. The FFA is on the lower surface of the brain. It is similar in principle to the superior temporal face area discussed in Chapter 4. There are indications that each performs slightly different functions, such as reading facial identity or deciphering emotion from facial expressions.

8

The Feeling of What Happened

At this point, mindful of our initial distinction between the "inspirational" and "instrumental" uses of cognitive and neuroscientific ideas, it is useful to turn to some "field studies" of how the former might emerge in an ever-changing fictional landscape.

Ian McEwan's *Saturday* (2005a) can be seen as the epitome of an "inspirational" use of neuroscience in literature. But McEwan's reach-out across the "two cultures" in this novel should come as no surprise. As early as 1987 in *The Child in Time*, he has a physicist called Thelma declare: "Shakespeare would have grasped wave functions, Donne would have understood complementarity and relative time" (McEwan 1987: 45). By the time of *Enduring Love* (1997), he is engaging not just with the implications of science *per se*, but directly with neuroscience. Marco Roth writes that *Enduring Love* "effectively inaugurates the genre of the neuronovel," and Andrew Gaedtke calls the novel "the most formally ambitious example of contemporary literature's engagement with cognitive science" (Owen 2017: 140).

About half-way through *Saturday,* the protagonist Henry Perowne, a successful neurosurgeon, is introduced to the British prime minister of the time. In the ensuing exchange of pleasantries, it becomes evident that Tony Blair has mistaken him for someone else, a painter. Slightly embarrassed, Perowne tries to correct the misconception—but the premier persists: "'In fact, we've got two of your pictures hanging in Downing Street. Cherie and I adore them.' 'No, no', Perowne said. 'Yes, yes', the Prime Minister insisted" (McEwan 2005a: 146). It's a moment of comic misunderstanding which is entirely appropriate given the focus of the novel within which it occurs: acts of self-deception, confabulation, and retrospective justification all falling within the author's brief. It also reminds us that some groups of people—here, politicians—seem impervious to certain social emotions. The incident has a striking air of authenticity—and indeed in an interview with Jasper Gerard, McEwan confirmed its autobiographical source (*Sunday Times*, January 23, 2003).

Our ability to convince ourselves that we are correct in spite of all evidence to the contrary—whether on the relatively trivial matter of someone's identity, or the decision to prosecute a war which will blight thousands of lives—is a key issue in *Saturday*, because it is a key issue in the examination of a character's subjective world. Indeed, with a nod to E. L. Doctorow (Chapter 10), a more accurate title for the book might have been *Henry's Brain*.

This is a novel rich with ideas derived from brain science—on evolution, morality, emotion, and feeling, deliberately taking us into many byways of the self, and our newly imperiled sense of agency. But even though McEwan readily takes up a neuroscientific map to complete Henry's journey, he also muses on its deficiencies and gaps, the features it fails to include, "who will ever find a morality, an ethics down among the enzymes and amino acids?" (2005a: 92).

The novel's protagonist is an uncompromising materialist, who doesn't so much ignore the Cartesian elephant in the room, as deny that it even exists: "It isn't an article of faith with him, he knows it for a quotidian fact, the mind is what the brain, mere matter, performs" (2005a: 66). Yet, McEwan reminds us there is nothing "quotidian" in matters of mind, as the events of Perowne's "ordinary" *Saturday* are about to prove. Henry's no-nonsense reductionism is subjected to serious strains in the course of the novel, whether confronted by his daughter's poetic sensibility, the ineffable beauty of his son's music, or his own emotional responses to acts of violence and intrusion. For Henry, despite being girded by professional certainties, an intimate knowledge of the brain, clinical, financial, and personal success (his surname indeed, an anagram of "peer won") is just as much a mystery to himself as we all are, his judgments, perceptions, and feelings less assured than he would care to admit.

This is evident in the wonderful opening sequence of the novel, in which he wakes in the dark "some hours before dawn," walks to the window, and witnesses in the night sky—what? A comet? A meteor? In many ways this exterior event, in all its shifting interpretations and associations, merely reflects the uncertainties of introspection. The event is a correlative of his own state, which is only tenuously known to him. We are, McEwan seems to be saying, not only adrift in the world, but adrift within ourselves. "He has no idea what he's doing out of bed: he has no need to relieve himself, nor is he disturbed by a dream or some element of the day before. It's as if, standing there in the darkness, he's materialised out of nothing, fully formed, unencumbered" (2005a: 1).

The sinuous surface of the prose, its effortless ease, reflects Henry's calm, but also points up just how estranged he is here from his waking, scientific, and decisive self. In this semi-somnolent state, his conscious being seems a flickering, uncertain thing, definable only negatively. He does not willingly rise from the bed, but "find(s) himself already in motion … He has no idea what he's doing out of bed." His awareness is defined by what it cannot do—by its limitations. For the duration of this nocturnal incident, Henry is obliged to hazard retrospective guesses about what it is "he" (his body) has elected to do, and why.

For all his apparent ease, he is reduced to making a series of hesitant deductions (stabs in the dark) about his own mental and physical condition. McEwan here creates an initial impression of Henry as someone trying to catch up with himself—and failing. Standing at the window, "He doesn't immediately understand what he sees, though he thinks he does. In this first moment, in his eagerness and curiosity, he assumes proportions on a planetary scale; it's a meteor burning out in the London sky, traversing left to right, low on the horizon, though well clear of the tallest buildings" (12).

The incident is also a pertinent example of some of the imaginative visual techniques (Chapter 2) used by Elaine Scarry when she asked us to imagine a dark featureless space, then to introduce a pinprick of light, making it streak across the blackness, as if it were moving across the retina: "Imagine this dark space with sudden flares and lights—bursting, then disappearing, streaming across the field of mental vision or arcing through" (Scarry 2001: 77).

By inviting us to share Henry's wavering, inchoate perception, the author is, at this very moment, triggering the emergence of our own shadow percepts, that is, instructing our imagination, via our memories. As readers, we travel *pari passu* with Henry. He is trying to see, and we, too, are (inwardly) trying to see. But what? Scarry would recognize what McEwan does next as a classic way to confer dimensionality on the default thinness— the "tissuelike" quality—of any imaginary scene. McEwan violently pulls the focus, forcing us to shift scale—first outward/upward and then inward/ downward. "You see [comets] in a flash before their heat consumes them. This [image] is moving slowly, majestically even. In an instant, he revises his perspective outward to the scale of the solar system, this object is not hundreds, but millions of miles distant" (12). It is the intrusion of another sense—hearing—that obliges him (and so, us) to reframe entirely what is being observed. Hearing a "low rumble" he realizes the object in the sky has to be much closer than he's thought. His inner mapping of the event requires urgent revision. As a reader, this plays out as an instance of the "inner ear" recalibrating and refining the workings of the "inner eye," and is a demonstration of the intimate relation between perception and memory.

Now that Henry is able to contextualize what it is he's seeing, he decides it's a plane: his brief satisfaction at resolving the scale of this new cognitive map dispelled by an unexpected consequence. If it's a plane, then a fire must have broken out on board. Now there are new gaps in the chart to be narratively completed. "The leading edge of the fire is a flattened white sphere which trails away in a cone of yellow and red, less like a meteor or comet than an artist's lurid impression of one." As well as being a prime example of using Scarry's techniques on us, this goes a radical step further to imply that, fundamentally, the act of seeing, and comprehending what we see, involves a degree of compositional control—that we are all the "artists" of our passing impressions. It is one of those "small spatial stories" of Mark Turner, in the process of being formed (Chapter 4).

The Pinprick of the Present

Perowne's unexplained waking—and what ensues from it—is a statement of intent by McEwan.[1] It shows him marking out his territory: introspection and its limitations. Faced with such an elusive quarry, McEwan constructs a plot replete with all the conceptual traps and snares his goal demands (Steadman 2008).[2]

That plot is this: Perowne, by way of a minor car accident, finds himself involved with a petty criminal—Baxter—who, near the end of the single day of the action, breaks into the surgeon's house, attacks his family, threatening to rape his daughter. Baxter is suffering from Huntington's disease, a chronic condition which produces violent mood swings. By a quirk of fate, after Baxter is finally disarmed, it is Perowne who ends up performing emergency brain surgery on him.

Here, narrative feasibility has clearly been sacrificed: fidelity to the neuroscientific theme trumping any pretense at verisimilitude. Indeed, Dominic Head describes *Saturday* ... as attempting to "'produce, perhaps,' a diagnostic 'slice-of-mind' novel—working towards the literary equivalent of a CT scan—rather than a modernist 'slice-of-life' novel" (Owen: 154). Which may be so, but we must also bear in mind that "a story is a kind of map and maps are created through the use of projections, or distortion models" (Turchi: 167).

In an article written by McEwan in *The Guardian* newspaper (April 1, 2006) about the "scientific literary tradition," he lists fifteen canonical works. They have a heavily biological slant, including Darwin's *The Expression of Emotions in Man and Animals*, Dawkin's *Selfish Gene*, and Ernst Mayr's *This Is Biology*. He also adds "Antonio Damasio's hypnotic account of the neuroscience of emotions in *The Feeling of What Happens*."

At first glance, this inclusion might seem a little incongruous, sitting as it does alongside Francis Bacon's *Advancement of Learning*, and Galileo's *Dialogues*. Yet Damasio's theorizing on consciousness as an emergent phenomenon has had a massive effect on McEwan, perhaps most visible in *Saturday*, leading McEwan into some of his most searching and forensic prose.

A good starting point is the description of a game of squash about 100 pages into McEwan's novel. Henry arrives late, delayed because of the car crash and subsequent run-in with Baxter. He is somewhat distracted, but determined to see the game through. His opponent is an anesthetist—Jay Strauss, "a powerful, earthbound, stocky man, physically affectionate, energetic, direct in manner" (2005a: 102).

Everything about Strauss—his bluntness, tendency to physical affection, and bulk—indicates that he is here allied with the body, with corporeality, only reinforced by the fact that he spends his professional life suspending his patient's consciousness. Perowne, on the other hand, is adrift in his own thoughts: head-bound as opposed to earthbound.

McEwan, describing the squash game, crosses and re-crosses the borderline between reporting Henry's thoughts and presenting them in apparently "unmediated" free indirect form, by way of inner stream of consciousness.[3] Indeed, as Stedman and others have noted, by restricting the action to one day, McEwan is invoking two hyper-canonical stream-of-consciousness novels that do the same: Joyce's *Ulysses* and Woolf's *Mrs Dalloway*.

This process of smuggling the reader into and out of Henry's head is used often in some of the more significant passages in the novel. At this point, he is feeling physically vulnerable: "He feels his left knee creak as he stretches his hamstrings. When is it time to give up this game?" (2005a: 103). Repeatedly, we move from narrated third person into the realm of thought. Abruptly if subtly, the coordinates shift before we become aware of it. Or rather, the entire map of the narrated "bird's-eye view" is whipped away only to be replaced by the partial and shifting inner schemata of the character. During the game, Perowne lags behind his own body, his mind at a third remove, speculating, questioning, and of course, given this, the match does not go well: "The ball surprises him—it's as if he left the court for a moment" (2005a: 107).

Memories and impressions from earlier in the day possess him, leaving the second-by-second demands of the game to be managed on "autopilot" (implicit memory). So great is his distraction that those other, autobiographical memories begin to usurp his sense of the present, "occupy(ing) the wrong time coordinates."

The Appearance of an Owner

Since McEwan himself provides the evidence, here and in his interviews, of the deep influence of Damasio, it is worth briefly revisiting some of the distinctions made by Damasio on the subject of consciousness. He begins by articulating what he sees as the two central problems lying behind our "feeling of what happens," the first being "the problem of how we get a 'movie-in-the-brain' ... how the brain makes neural patterns in its nerve-cell circuits and manages to turn those neural patterns into ... images" (Damasio 2000: 9).

The second problem is related but quite distinct—how does "the brain ... engender a sense of self in the act of knowing." This second phenomenon—call it "proprietary knowledge"—is to do with how we create an owner and observer of the first.

Rejecting any metaphysical source for this witnessing "self," Damasio insists that it emerges naturally from neural processes. There is no "soul" or mysterious homunculus at work; the observing self must be created and apprehended in a similar way to other exterior and interior events. So, "the second problem is that of generating the *appearance* of an owner and observer for the movie within the movie" (2000: 11, his italics). The proprietary self arises from and partakes in the "movie-in-the-brain." And note that word—"appearance"—he is already putting traditional ideas of personal identity up for grabs.

Damasio's analysis entails distinct levels of awareness—as he repeatedly tells us, "consciousness is not a monolith" (2000: 121). "*Core consciousness* provides the organism with a sense of self about one moment—now—and about one place—here ... [it] does not illuminate the future, and the only past it vaguely lets us glimpse is that which occurred in the instant just before." It is a biological phenomenon not dependent on memory or language, and shared by other species. Its instantaneous, evanescent quality, though, makes it unable to furnish us with a stable long-term sense of self. *Extended consciousness*, on the other hand, is more durable, and distinctively human. "In extended consciousness both the past and the anticipated future are sensed along with the here and now in a sweeping vista as far ranging as that of an epic novel" (2000: 16–17).

Others have also turned to the heightened spatio-temporal awareness that occurs at key moments in sport to investigate the conscious mind. John McGrone opens his study: *Going Inside: A Tour around a Single Moment of Consciousness*, with the description of an instant that occurred to him during a netball game, "I had exploded into the air with such venom that it seemed to have stopped time. The world was moving with treacly slowness, illuminated with a giddy brilliance. I dangled above the net, the only one free to think and act ... It only happened once" (1999: 1). Yet this second or two, and the speculations that came from it—"Did time always freeze for really good players? ... Or was my moment only an illusion?"—led him into a long-term study of consciousness. Repeatedly, in *Saturday,* McEwan uses just such adrenaline-filled moments, like Henry's car crash, and the assault on his family, to allow us into this "second or two" where the appearance of "stopped time" brings core consciousness to the fore. But this exclusive focus on the granular detail of the "now" cannot last, and is soon to be swept up in the workings of the narrative self.

Within the context of *Saturday* and the episode on the squash court, McEwan gives us a voice which cleverly conflates Henry's inner thoughts and the exterior report of the narrator. Just as Henry is trapped in the hinterland between third and first person, so too, it appears, is his story. "The game becomes an extended metaphor of character defect. Every error he makes is so profoundly, so irritatingly typical of himself ... like a scar tissue, or some deformation in a private place. As intimate and self-evident as the feel of the tongue in his mouth" (2005a: 103). Is this McEwan writing about Henry in the third person, or Henry thinking about himself in the third person? Either way, the tendency to abstraction, "occupying the wrong time coordinates," is plain. In an insidious feedback loop, his errors on the court reinforce the self-lacerating ruminations of the "autobiographical self" that led to them in the first place.

There is a further irony here, in that McEwan resorts to corporeal similes to express Henry's (abstracted) self-loathing, "like a scar tissue or some deformation in a private place." The point is surely clear. There is no escape from the body. Henry can only be "headbound" to an extent—ultimately, we're all like his opponent, Strauss—tied to our bodies.

Later, after a break in play, he returns in a very different frame of mind, ready now to exist in the present, in the way the game demands. "It's possible in a long rally to become a virtually unconscious being, inhabiting the narrowest slice of the present" (111).

And, although it is made clear that the game is merely a tussle between two middle-aged players ("there's nothing at stake") at another level, the outcome of the match assumes huge importance, symbolizing as it does a more primal struggle, "the irreducible urge to win, as biological as thirst" (115).

McEwan places all the major events of the novel against an evolutionary backdrop. Indeed, the first mention of Darwin is as early as page four, where we see Perowne lolling in the bath. Not only is this apt from an evolutionary perspective, since life emerged from the sea, but lest we miss the point, he is also reading a biography of Darwin—the scientific figure most regularly invoked in McEwan's many interviews and articles:

There is one preeminent scientist who is almost as approachable ... as a novelist. It is perfectly possible for the nonscientist to understand what it is in Darwin's work that makes him unique and great. In part, it is the sequence of benign accidents that set him on his course ... And partly it is the subject itself. Natural history, or nineteenth-century biology generally, was a descriptive science.

(McEwan 2005b: 7)

One of the weaker aspects of *Saturday* is the way we often feel led by the nose like this. McEwan in the interview above, and also in many of his novels—*Enduring Love* (1997) *Saturday,* and, most recently, *Machines Like Me* (2019)—is so fixated with scientific concepts that he risks appearing clumsy or schematic. So, the Darwinian theme is reprised repeatedly. The opening line of chapter two, "There is grandeur in this view of life," a direct quotation from *Origin of Species,* echoes over and over through Henry's head. And, just to be sure, McEwan elaborates, "endless and beautiful forms of life, ... including exalted beings like ourselves, arose from physical laws, from the war of nature, famine and death. This is the grandeur. And a bracing kind of consolation in the brief privilege of consciousness" (2005a: 54).

The Feeling of What Happens is a relentless, detailed pursuit of what that "brief privilege" actually entails. In it, Damasio makes the case that core consciousness is an evolutionarily ancient phenomenon, "the sites [associated with] core consciousness are located near the brain's midline. These structures are of old evolutionary vintage, they are present in numerous non-human species" (Damasio 2000: 10).

By contrast, extended consciousness, drawing on memory and higher thought, is a stable, enduring sense of the self over time. This would explain why extended consciousness depends on the functioning of core consciousness but not vice versa, just as, he suggests, the workings of an "autobiographical self" depend on the functioning of a "core self" but not vice versa. In each case (as the names suggest), core consciousness and core self are primary. A host of lesion studies, clinical observations, and theoretical probabilities are marshaled in support of this contention. "Behind extended consciousness, at each and every moment, lies the pulse of core consciousness. This may sound surprising, but it need not be. We still need digestion to enjoy Bach" (2000: 125).

This last sentiment could have been transcribed directly from the pages of *Saturday*. Damasio explains that core consciousness is generated in "pulse-like fashion":

> It is the knowledge which materializes when you confront an object, construct a neural pattern for it, and discover automatically that ... the object is formed in your perspective, belongs to you, and you can even act on it. You come by this knowledge ... instantly: there is no noticeable process of inference, no out-in-the-daylight logical process that leads you there, and no words at all—there is the image of the thing, and, right next to it, the sensing of its perception by you. (126)

This goes to the very heart of the issues confronted professionally and personally by Henry Perowne. We can now begin to understand why he is depicted so often as being "in the dark" both literally and metaphorically— as in the novel's opening sequence. The genesis of "self" is essentially unfathomable, mysterious, inaccessible to either introspection or any "out-in-the-daylight" logical process. The means by which Henry's impressions become Henry's impressions are not themselves available to Henry.

Belated and Benighted

The most intimate and foundational process—that of first-person perceptual ownership—is itself inscrutable, according to Damasio. "What you do not ever come to know directly is the mechanism behind the discovery, the steps that need to take place behind the seemingly open stage of your mind in order for core consciousness of an object's image to arise and make the image yours" (126).[4] Worse, not only is the process of generating core consciousness itself hidden from us, but it takes time: time during which the process, and even the object, has changed.[5] "By the time you get 'delivery' of consciousness for a given object, things have been ticking away in the machinery of your brain for what would seem like an eternity to a molecule—if molecules could think. We are all living in the past, and because we all suffer from the same tardiness, no-one notices it" (Damasio 2000: 126). Even when Henry tries to be truly present, he is doomed to fail. He is forever, as Damasio says we all are, "hopelessly late for consciousness."

This fundamental disjunction is only compounded by the tendency for memory to hijack perception. Take the pivotal moment of Henry's day—his car crash with Baxter. In the lead-up Henry's thoughts do not relate to his bodily actions (the physical reality of driving a car) but rove over a whole series of musings about his own emotional tone. "He's still bothered by his peculiar state of mind .... the proximal and distal causes of his emotional state. A second can be a long time in introspection" (78).

Ever the scientist, Henry attempts to anatomize the causes "proximal and distal" of his current mood. It is this moment of failed introspection which precipitates the car crash, thus Baxter's vendetta and all that ensues.

> The assertions and questions don't spell themselves out. He experiences them more as a mental shrug followed by an interrogative pulse. This is the pre-verbal language that linguists call mentalese. So that when a flash of red streaks in across his left peripheral vision, like a shape on his retina in a bout of insomnia, it already has the quality of an idea, a new idea, unexpected and dangerous, but entirely his, and not of the world beyond himself (81).

Preoccupied, stubbornly assaying his emotions, he remains at the mercy of his crude instruments—words and concepts, questions and answers. But, disobligingly, they "don't spell themselves out .... He experiences them more as a mental shrug followed by an interrogative pulse." Given McEwan's many allusions to Damasio, it is hard not to think here of the "pulse of core consciousness." "The continuity of consciousness ... based on the steady generation of consciousness pulses" (Damasio 2000: 176).

McEwan is aware he is leading us into deep waters, unfathomable with language alone, undertaking a linguistic enterprise that is at the very least fraught, precisely because the phenomenon it attempts to reach is pre-linguistic (see Porter Abbott 2013, for a discussion of this recurrent paradox).

The Poetic Present

Yet McEwan persists in the attempt and even offers the slender hope that it just might be possible. And how? Despite everything: through poetry. Poetry plays a vital role in the plot of *Saturday*. Both Henry's daughter, Daisy, and father-in-law (the clunkily named "Grammaticus") are poets.

Seeking vengeance, intent on rape and murder, Baxter breaks into Henry's house, forcing Daisy to strip at knifepoint. Yet, at this crux moment, it is a poem of all things that abruptly transforms the mood of the would-be rapist, entrancing and disarming him. Why poetry? Characteristically, McEwan spells it out for us. "Novels and movies ... propel you forwards or backwards through time, through days, years, or even generations. But to do its noticing and judging, poetry balances itself on the pinprick of the moment" (2005a: 129).

Is it a coincidence that McEwan describes the vision of the burning plane in chapter one as a "pinprick of fire" and here deploys precisely the same simile to describe the present? It underscores the point that the episode of the plane is an instantiation of "hereness," "nowness," and one which Henry fails to grasp, possessed as he is by the demons of over-interpretation (107).

In *The Feeling of What Happens*, Damasio, in his efforts to articulate our perception of core consciousness, also turns to poetry and music. Speaking of our "primordial narratives of consciousness," he elaborates:

> I do not mean narrative or story in the sense of putting together words or signs in phrases or sentences. I do mean telling a narrative or story in the sense of creating a nonlanguaged map of logically related events. ... A line from a poem by John Ashbery captures the idea: "This is the tune, but there are no words, the words are only speculation."
>
> (Damasio 2000: 185)

Setting aside the paradox of using words to transcend language, the parallels to what McEwan is doing are striking.

So, even though words cannot capture the immediacy of core consciousness in and of itself, Damasio suggests that our need to tell stories, to write plays and poetry derives from the very nature of this "foundational" awareness. There is a quality inherent in the generation of core consciousness which predisposes us to tell "stories" in the widest sense. In this regard, a poem is yet another "story," but one of peculiar and focused specificity, "balanced on the pinprick of the moment."

Damasio pursues the point: "Wordless storytelling is natural. The imagistic representation of sequences of brain events ... is the stuff of which stories are made. A natural preverbal occurrence of storytelling may well be the reason why we ended up creating drama and eventually books" (188).

This, as we saw in Chapter 1, is the "narrativist" claim writ large. Mark Turner tells us that perception and thought are based on "small spatial stories" (Chapter 4). And Damasio concurs, "Telling stories precedes language, since it is, in fact, a condition for language" (189).

In *Saturday*, the question of what "stories" (poems, plays, films) are actually "for" underlies almost every scene. As a scientist, Henry is skeptical, "fiction is too humanly flawed, too sprawling and hit-and-miss to inspire uncomplicated wonder at the magnificence of human ingenuity" (McEwan 2005a: 67).

Despite his fundamental distrust of fiction as an analytical tool, Perowne dutifully trudges through reading lists (*Anna Karenina*, *Madame Bovary*) given him by Daisy, but finds them "merely the products of steady, workmanlike accumulation" (2005a: 66). The choice of these particular novels is interesting because they are the very ones McEwan cites in his essay "Literature, Science, and Human Nature," as outstanding instances of how the best literature taps into our "mentalizing" skills:

> By an unspoken contract, a kind of agreement between writer and reader, it is assumed that however strange these people are, we will understand them readily enough to be able to appreciate their strangeness. To do this, we must bring our own general understanding of what it means to be a person. We have, in the terms of cognitive psychology, a theory of mind, a more-or-less automatic understanding of what it means to be someone else.
>
> (McEwan 2005b: 5)

Perowne is, it seems, closed to such a "contract," deaf to the empathetic invite behind it.

Even the poetry of his daughter and father-in-law he struggles to enjoy, failing to see the point of such condensed effort. Finally, in exasperation he concludes: "This notion of Daisy's, that people can't 'live' without stories, is simply not true. He is living proof" (2005a: 67).

But of course, the edifice of *Saturday* contradicts Henry. It is its own "living proof" that he is wrong. Setting aside the obvious point that, *qua* character, he only exists within the bounds of this particular story, above and beyond this, the narrative of the novel, composed as it is of his own shifting ruminations and interpretations of the world and of himself, is an ongoing refutation of this statement. Henry would like to believe, like philosopher Galen Strawson (Chapter 9) that he is "not a story," but he is actually an excellent example to the contrary. Despite himself, he endlessly spins and is spun by his stories.

From the opening nocturnal sequence, where Henry shows himself, as we all are, prone to misinterpretation, imaginative extrapolation, to his run-in with Baxter, to the torments of his squash game, and to the final confrontation in his home, Henry reveals himself to be (despite all his protestations) an indefatigable storyteller, a "scientific" Scheherazade.

Possible Futures

You might say it is precisely Henry's need to make sense of what he sees—to narrativize it—which leads him astray. As in the incident with the plane, impelled by a *horror vacui*, Henry repeatedly uses story to paper over the cracks in his knowledge, distorting the evidence before him in the process, his inner map filling in the voids, misrepresenting incongruent or inconvenient perceptions.[6]

So, in the second his car hits Baxter's we hear that, "when a flash of red streaks in across his left peripheral vision, like a shape on his retina in a bout of insomnia, it already has the quality of an idea, a new idea, unexpected and dangerous, but entirely his, and not of the world outside himself" (2005a: 81). The allusion to the incident with the plane, "a shape on his retina," is telling. Here too, faced with imminent danger, his brain re-casts reality, in this case co-opting the "flash of red" into his own mental narrative, making it "entirely his" in a bid to transform a worldly threat into an imagined one: again, in the deepest sense, deploying story—"a new idea"—to allay fear. Under extreme pressure, he momentarily conflates "self" and "other," the subjective and the objective. Henry, instead of living his life, is fictionalizing it. He has become his own reader.

Even when Baxter breaks into his house and threatens the lives of his family, Henry is unable to free himself from the hypnotic spell of stories—this time in the shape of "possible futures"—redemptive but, given the situation, enervating narratives: "In his usual manner he's been dreaming—of 'rushing' Baxter with his son Theo, of pepper sprays, clubs, cleavers, all stuff of fantasy. The truth, now demonstrated, is that Baxter is a special case—a man who believes he has no future and is therefore free of consequences" (2005a: 217). The suggestion is that Baxter's problem is the opposite of Henry's. Baxter because of his neurological condition is trapped in the instant, unable to contextualize his situation within the wider frame of "autobiographical self": his self-story unraveling. "It's of the essence of a degenerating mind, periodically to lose all sense of a continuous self, and therefore, any regard for what others think of your lack of continuity … he inhabits the confining bright spotlight of the present." Little wonder, then, that it should be poetry, expressing the "pinprick of the moment" (2005a: 232, 129) that should move and pacify him.

As seen, for Damasio, there is a convincing case for the emergence of consciousness from the homeostatic systems of the body. He lays out his argument: core consciousness does not depend on "conventional learning and memory," or language. However, "emotion and core consciousness are clearly associated" (2000: 232). So, in a patient whose core consciousness has been impaired, the expression of emotion is absent. And emotion (others might say core affect) has to be regarded as a primary homeostatic process monitoring and acting on the body as a whole. "At their most basic, emotions are part of homeostatic regulation, and are poised to avoid the loss of integrity that is a harbinger of death" (2000: 54).

Emotions and derived feelings are part of a largely preset neural system, Damasio says, which sends commands to other parts of the brain and to the body, via the bloodstream and neuronal electrochemical signals, resulting in a "global change in the state of the organism" (2000: 67).

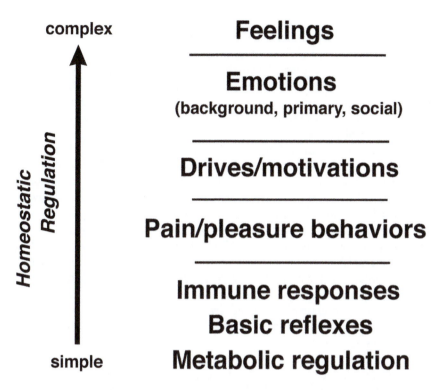

FIGURE 8.1 *Antonio Damasio's model relating emotions and feelings to broader homeostatic mechanisms.*

Yet much of this action is unconscious (the lower tiers in Figure 8.1). Two further steps must take place before an emotion is known. The first is the imaging of these very same body changes, that is to say: feeling. The second is the application of core consciousness to the entire set of phenomena. "Knowing an emotion—feeling a feeling—only occurs at that point" (2000: 68).

Henry sits in his car in the aftermath of the crash, listening to the tick of cooling metal, a moment McEwan uses to foreground the analogy between car and driver, body and brain,[7] just as, later in the novel, he will explicitly invoke the extended metaphor of the map of London as one massive cerebrum—the dendritic streets teeming with synaptic "traffic."

Henry tries to take in the extent of the damage incurred in the crash—which, as it turns out, is confined to the "bodywork." "His car will never be the same again. It's ruinously altered, and so is his Saturday … there swells in him a peculiarly modern emotion—the motorist's rectitude, spot-welding a passion for justice to the thrill of hatred" (2005a: 82).

At this point, McEwan pulls a trick we've seen before, eliding subject and object, so that in a sleight of hand, we shift from "the car" to "Henry," or, at any rate "Henry's Saturday." "It's ruinously altered." Then, giving the metaphor a cognitive slant, McEwan adroitly superimposes the idiom of auto-repairs onto the inner workings of Henry's brain: "spot-welding a passion for justice to the thrill of hatred."

To step out of the plot of *Saturday* for a moment, McEwan as writer is certainly alive here to another quite distinct use of the word "motor," in a context Henry would have well understood. Given the neuroscientific themes addressed in the novel, it's one of the reasons the motor accident is such a powerful device.

As previously outlined, functional brain imaging has shown that the act of reading activates parts of the nervous system not only in sites dealing with semantics and syntax, but often in sensory and motor areas relating to bodily sensation and behavior *associated with the words that are read*. So action words (kick, lick) sometimes activate the motor cortex, reading "cinnamon" or "garlic" may activate the olfactory cortex, and textural metaphors can activate the somatosensory cortex (Hauk et al. 2004, Gonzalez et al. 2006, Lacey et al. 2012). If this is the case, then reading—as we are about Henry's accident—may involve more than abstract operations like checking an inner lexicon (semantic memory), and extend into sensory and motor (muscular) reenactments—simulations—evoked by them: a strong interpretation of embodiment (see Tettamanti et al. 2005, Pulvermüller and Fadiga 2010, Esopenko et al. 2012).

Others might take the position that if the brain-mediated flow of information between our mind and the world of our sensory and motor experiences is so substantial, then the environment might actually be considered part of our cognitive system. The maps are all part of one atlas.

However, the best way to look at this might be to say that "situated" or embodied cognition is not an all-or-nothing proposition (Wilson 2002). It may apply to some forms of cognition, but not universally. (For a thoughtful extended discussion, see Arbib et al. 2014.) In sum, situated cognition is an intriguing idea that challenges us to better understand the way sensory and motor systems relate to cognition in our daily lives (see Tranel et al. 2001, 2003; for a case where they don't, see Vannuscorps and Caramazza 2016).

In the accident scene, as elsewhere, the ultra-rationalist Henry, in the act of invoking his scientific heritage, is above all subject to his feelings, whether he likes it or not. Just like the rest of us, he is in thrall to "the feeling of what happens."

Down among the Wet Stuff

Lest we forget that all of these levels of analysis, description, and introspection ultimately depend on the "wet stuff," the three-and-a-half-pound mass in our heads, toward the end of the day, Henry finds himself staring down into the brain of the man who had tried to destroy him: Baxter.

Peering at the familiar pink-gray tissue, Perowne wonders, again turning to the analogy of film:

That mere wet stuff can make this bright inward cinema of thought, of sight and sound and touch bound into a vivid illusion of an instantaneous present, with a self, another brightly wrought illusion, hovering like a ghost at its centre. Could it ever be explained, how matter becomes conscious? He can't begin to imagine a satisfactory account, but he knows it will come, the secret will be revealed. (2005a: 262)

Indeed, "That's the only kind of faith he has. There's grandeur in this view of life" (2005a: 263). The Darwinian clarion call sounds once more, this time in defense of scientific method.

The scene in the operating theater is an important moment, both as regards the plot (will Perowne take justice into his own hands?) and also because it goes to the root of all the questions McEwan has been raising. Baxter's "wet stuff" is, in a sense, the source of all the viciousness of the day, and is now revealed in all its awful vulnerability.

We are presented with the same question Henry is—given his condition, can Baxter be held responsible for such behavior? Just a few hours before, facing the intruder, we heard that, despite his materialism, "Perowne can't convince himself that molecules and faulty genes alone are terrorizing his family and have broken his father in law's nose" (2005a: 218). Or, again, translated into philosophical terms, "who will ever find a morality, an ethics, down among the enzymes and amino acids?" (92).

And at this decisive moment, with Baxter's life in his hands, and his years of expertise coming to the fore, there is a clarity, a sharpness about Henry's thinking, his sense of awe quickly followed by a hard-edged assertion that our sense of "self"—the Cartesian *ne plus ultra*—is actually insubstantial. Typically, McEwan drops this bombshell in a parenthetical aside: "a vivid illusion of an instantaneous present, with a self, another brightly wrought illusion, hovering like a ghost at its centre" (262). Not only is the present a "vivid illusion," but so too, in a memorable image, is "the self," which we are told hovers like a ghost (in the machine) at its center. One mirage nests inside another.

Damasio too realizes he has to confront this highly charged issue: personal identity. To do so, he introduces another entity—the "*proto-self*": "I propose that the sense of self has a preconscious biological precedent, the proto-self, and that the earliest and simplest manifestations of self emerge when the mechanism which generates core consciousness operates on that non-conscious precursor. The proto-self is a coherent collection of neural patterns, which map, moment by moment, the physical structure of the organism in its many dimensions" (Damasio 2000: 153). Operating far beneath the threshold at which we are conscious, it monitors the state of the organism at multiple levels.

For him, the presence of "core self" is dependent on "proto-self," just as the more extended "autobiographical self" is dependent on "core self." It might be simplest to explain the proto-self as the most basic inner representation of the organism at any one time. So what occurs to make the unconscious proto-self cross the boundaries of awareness and become core self? Change. Interaction. The proposed causal chain is this:

1   The organism interacts with an object.

2   Both organism and object are mapped in the brain.

3   Then, "the sensorimotor maps pertaining to the object cause changes in the maps pertaining to the organism" (2000: 169).

4   These changes can be *re-represented* in higher "second-order maps," holding the relationship between object and organism. The mental images that describe this relationship (being body-related) are feelings.

Most importantly, "As far as the brain is concerned, the organism in the hypothesis is represented by the proto-self" (2000: 70).

So, the brain forms images of an object—and we are talking about "objects" in the widest possible context—it could be a face, a poem, a memory, a pain in your toe. Two neural accounts are initiated—the first details the changes occasioned in the proto-self by the encounter with the object. The second ("second-order maps") re-represent this as an account of a "causal relationship." This is the crux, "one might say that the swift, second-order nonverbal account narrates a story: that of the organism caught in the act of representing its own changing state as it goes about representing something else" (2000: 170). The process behind the generation of "self," then, might be described as not merely narrativism, but metanarrativism.

At first glance, this seems startling enough, but then Damasio spells it out: "The astonishing fact is that the knowable entity of the catcher has just been created in the narrative of the catching process."

In other words, the sense of self in its earliest conscious stages arises from a re-representation (a narrative) concerning the (non-conscious) proto-self and its modifications. Thus, the "narrator" emerges naturally from and is represented in the organism's "narrative." "The images in the conscious narrative flow like shadows along with the images of the object for which they are providing a ... comment. To come back to the metaphor of the movie-in-the-brain, they are within the movie. There is no external spectator" (2000: 171). Shockingly, perhaps, what Damasio is suggesting is that the "self" is a character emergent from a self-generated story. The very process of creating "higher-order" maps of experience not only locates, but also *generates the entity* of the mapmaker. Or, put another way, "I" am both author and character. Certainly, there are moments in McEwan's novel, like the car crash and the airplane, where Henry comes tantalizingly close to realizing this very thing.

Deckchairs, Coat-hangers and the Music of the Self

We may now be approaching the reason why McEwan elects to use a flexible third-person narrative voice that, periodically, shades into an intimate first person. It makes sense precisely because Henry is both "catcher" and "caught," dancer and dance, narrator and narrated—the "third-person" witness to his own thoughts and deeds, *and* the first-person subject of them. Far from being able to "live without stories," he is enwrapped, entranced, and entangled in stories at every level—narratives which "flow like shadows" alongside his every interaction are indeed inescapable, because they give rise to the self who is both producer and consumer of the "movie-in-the-brain."[8]

Yet, ironically, Henry does not see himself as creative. On the one occasion when we see him writing medical reports, McEwan uses the episode to hammer home the links between the mental and the physical. "His prose accumulated awkwardly. Individual words brought to mind unwieldy objects—bicycles, deckchairs, coat-hangers—strewn across his path. He composed a sentence in his head, then lost it on the page, or typed himself into a grammatical cul-de-sac and had to sweat his way out" (2005a: 11).

McEwan the novelist has fun here—watching Henry fighting his way through the "bicycles, deckchairs, and coat-hangers" of his own inarticulacy—but the passage is also a pointed reminder of the primacy of the physical. And, even when Henry tries to work out the reasons for his "debility" at this moment, he can't decide whether it's the "cause or … consequence of fatigue"—body and mind indissolubly linked, as always.

Damasio, in the course of his account, resorts to many artistic metaphors to express the ways in which consciousness is generated. The most fundamental, as seen, is that of "story," which for him includes poetry. In *Saturday*, the moment of greatest violence, with Baxter pointing a knife at a naked Daisy, is unexpectedly defused when she reads out Matthew Arnold's poem, *Dover Beach* ([1851] 2005a: 229). Hearing *Dover Beach*, Baxter is, we are to believe, transported, overtaken by a humbling sense of wonder. It is another thematically driven choice by McEwan, since Arnold's apologia makes the case for literature as a serious pursuit faced with the achievements and breakthroughs of science, a humanist rallying cry and a counter to a materialist and reductionist methodology which would ultimately leave us abandoned on the "darkling plain" of a Godless universe. It also carries echoes of a much later debate we've mentioned: between C. P. Snow and F. R. Leavis.

Arnold later lays out his humanist stall in his essay "Literature and Science" (1882) in response to a speech by "Darwin's bulldog" T. H. Huxley, whose stark empiricism in his own essay "Science and Culture," as we've seen, gave early rumblings of the two cultures debate that McEwan is very much alive to. "As with Greek, so with letters generally: they will some day come, we may hope, to be studied more rationally, but they will not lose their place ... we shall all have to acquaint ourselves with the great results reached by modern science, yet the majority of men will always require humane letters" (14).[9]

This speech both anticipates the current "instrumental" use of the brain sciences in literary studies and expresses some of the insecurities about the diminished role of the narrative arts, which permeates the work of those—like McEwan and Doctorow (Chapter 10) who enter this arena. But the underlying plea, for literature to be "studied more rationally," would certainly be one that Richards would have fully endorsed.

And although Henry sees himself as immune to the facile entrancements of storytelling, we do see him undergo his own moment of transport, under the sway of his son's music (2005a: 28). As he is obliged to admit, music is "a form of hypnosis, of effortless seduction" (28). Much later, we see Henry entering into an analogous state of "flow" as he exercises his professional skills operating on Baxter—hours during which he feels "calm and spacious, fully qualified to exist" (226).

At the very end of his tale, of his valuable Saturday, Perowne has become uneasily apprised of a deficiency in himself, a truth which finally becomes undeniable at the moment Daisy recites *Dover Beach*. "Baxter fell for the magic, he was transfixed by it .... But Baxter heard what Henry never has, and probably never will" (228).

With regard to the scientific paradigms behind *Saturday,* the magic of poetry and music is central to Damasio's argument. When trying to express how "the conscious you" comes about, he repeatedly has to turn to poetry, and, through it, music.

T. S. Eliot might have been thinking of the process I just described when he wrote, in the *Four Quartets*, of "... music heard so deeply that it is not heard at all," and when he said "... you are the music while the music lasts." He was at least thinking of the fleeting moment in which a deep knowledge can emerge—a union, or incarnation, as he called it (Damasio 2000: 172).

The "magic" that transfixes Baxter and Daisy, and Theo and even Henry, exhibits its power because it is fundamental to our sense of ourselves. It accompanies us as the "wordless story of self" that is the well-spring of all music, all poetry, all narrative:

> The story [is not] really told by you as a self because the core you is only born as the story is told, within the story itself. You exist as a mental being when primordial stories are being told, and only then; as long as primordial stories are being told, and only then. You are the music, while the music lasts. (2000: 191)

To Henry, tenderly kissing his wife's neck, quite suddenly, the far horizons and grand scale of his autobiographical self, a temporal map crammed with proliferating and importunate details of possibility and regret, collapse, giving way to a more miniscule framework, one which is perhaps as close as we can come to the "pinprick of the present." At the end of his long day, "There's always this ... And then: there's only this" (2005a: 289).

Notes

1   McEwan has a fondness for laying out his stall early in his novels—see for instance the breathtakingly vivid opening to *Enduring Love* (1997) (revolving around the delusional brain condition de Clerambault's syndrome) with its starkly visualized moral dilemma of an out-of-control helium balloon with a ten-year-old boy on board.

2   Some critics have attacked what they see as a pattern of overly schematic, even formulaic novels on scientific themes. "To write a novel which will be successful in terms of readers, reviews, and prizes, you need the following basic ingredients: an expert on neuroscience, for instance a neurologist; a victim or patient, for instance someone with a brain impairment following an accident; the victim's or doctor's wife, sister or girlfriend; a generally threatening, gloomy context, preferably something to do with 9/11... In other words: a brain plot, a sure recipe for literary success." Gesa Steadman, Brain plots: Neuroscience and the contemporary novel (2008) *Yearbook of Research in English and American Literature* 24:113–24 periodicals.narr.de/index.php/real/article/download/1615/1594.

3   There is an excellent analysis of both this narrative blurring in *Saturday* and its treatment of the theme of time, in Janice Rogers, *Unified Fields* (2014), chapter 8.

4   Ibid., p. 126, see also Christof Koch, *Consciousness* (2012) for an accessible account of current neuroscientific thinking on consciousness.

5   "we are rather in the position of Tantalus, with this difference, that we allow ourselves to be tantalised ... the aspirations of yesterday were valid for yesterday's ego, not for today's. We are disappointed at the nullity of what we are pleased to call attainment. But what is attainment? The identification of the subject with the object of desire. The subject has died—and perhaps many times—on the way." Beckett, *Proust* (1999) p. 14. See also Damasio p. 217: "The paradox identified by William James—that the self in our stream of consciousness changes continuously as it moves forward in time, even as we retain a sense that the self remains the same while our existence continues. The solution comes from the fact that the seemingly changing self and the seemingly permanent self, though closely related, are not one entity, but two. The ever-changing self-identified by James is the core self..... The sense of self that appears to remain the same is the autobiographical self, because it is based on a repository of memories" McEwan cites William James (*Saturday* p. 56) on the related issue of "forgetfulness."

6   Nassim Nicholas Taleb, on what he dubs "the narrative fallacy"—"We like stories, we like to summarize, and we like to simplify...The narrative fallacy addresses our limited ability to look at sequences of facts without weaving an explanation into them, or, equivalently, forcing a logical link, an arrow of relationship, upon them," p. 64 Nasim Nicholas Taleb, *The Black Swan* (2008).

With a nod and a wink, McEwan has Henry Perowne meet with a "Professor Taleb" in an Iraqi restaurant near Hoxton (*Saturday*, p. 287).

7   A parallel can be drawn here with the dystopian visions of J. G. Ballard, some of which explore the same territory, in particular, *Crash* (1973).

8   Whereas novelists can easily slip between first-, second-, and third-person perspectives in narrating cognitive states, neuroscientists have been trying to analytically link two disparate perspectives—the first person and third, for some time, see, for example, Northoff and Heinzel (2006). "Since mental states are experienced in first-person perspective and neuronal states are observed in third-person perspective, a special method must be developed for linking both states and their perspectives.....First-Person Neuroscience uses methods for the systematic examination and evaluation of mental states.... as experienced in first-person perspective and links them with data about neuronal states as obtained in third-person perspective." There is clearly some way to go.

9   Full text, http://homes.chass.utoronto.ca/~ian/arnold.htm, 3, speech first delivered on October 1, 1880.

9

Memory, Imagination, Self

In her novel *The Mind-Body Problem* ([1983] 1993), Rebecca Goldstein gives an exceptionally apt and humorous view of one woman's scheme through memory and imagination to chart her social world. In an aside on her subjective way of tracking the tense academic environment she was trying to navigate, she revealed a map that she invented, a "mattering map."

> We Princetonians live together on the mattering map ... constructed around a private image of a vast and floating map composed of untouching territories. Philosophers may prove the nonexistence of mental images; yet I don't think a week goes by that this one doesn't flash momentarily before me, called forth by someone saying something revelatory of his location in my private picture.

And she goes on from zones where people are categorized by external items, food, clothes, or music to:

> some intrinsic quality of his or her own: beauty or physical fitness. Or intelligence. Since we can discard these attributes even less easily than our clothes, we can always be strictly categorized according to the perceptions emanating from these areas: of who matters (the beautiful, the athletic, and the intelligent, respectively) and who doesn't (the ugly, the flabby, and the dumb).
>
> (22–3)

Memorializing our perceptions (and judgments) in memory and then calling them up repeatedly are a set of processes that we all use to navigate the physical world, and as above the social world. Narrative studies, brain science and philosophy all have much to say about such navigations and ultimately a sense of who we are.

The Territory of Mnemosyne

It's no exaggeration to say that memory is the foundation of our mental life and our sense of self, as I. A. Richards knew very well, "there is no kind of mental activity in which memory does not intervene" (1924: 97). Memory was, of course, the original repository of the great oral narratives within the storytelling tradition. Without it, the *Iliad* and the *Odyssey* would never have been passed down to us. As A. S. Byatt (2008) reminds us, "Mnemosyne, was, the Greeks believed, the mother of the muses" and narrative art is at its roots, "both a mnemonic and a form of memory" (xvi).

The literary and cultural historian Alan Richardson has documented how as the "doctors, philosophers, and proto-psychologists" of the eighteenth and early nineteenth centuries "changed the terrain for theorizing about the mind" toward materialism, literary culture was heavily involved (2001: 6). While Erasmus Darwin, Charles Bell, and others were making new observations and clarifying ideas (on sensation, bioelectricity, consciousness, and feelings), Coleridge, Keats, and George Eliot were actively contributing to the debates.

To take one example, a fundamental idea that can be traced to this era in Sir Charles Bell's writings is that the brain is not one unified "sensorium," but rather a collection of somewhat independent, or modular, units, marking the beginnings of a concept of functional differentiation, and anticipating our current acceptance of memory as a multifaceted process. It also opened up new ways to think about how memory is related to imagination. Coleridge generated a large number of suggestions on this very interface. He proposed that when we summon images stored in memory and recombine them we are engaging our powers of primary imagination or *fancy*, as he called it. This he distinguished from *secondary imagination*: the strongly creative power that "is capable of producing original ideas and images and [is] the shaping force behind all truly great works of art" (Richardson 2011: 282). Yet another instance of dialogue across "the two cultures," writers and poets equally contributing to the gestation of the sciences of mind.

Such dialogue is even more evident today. In E. L. Doctorow's novel *Andrew's Brain* (next chapter), about a "freakishly depressive cognitive scientist klutz" (2014a: 84), the central character muses on *The Prince and the Pauper*, where the two boys exchange identities. "It's more than a democratic parable: it's a tale for brain scientists. Given the inspiration, anyone can step into an identity because the brain is deft, it can file itself away in an instant. It may be stamped with selfhood, but let the neurons start firing, and Bob's your uncle" (105). Our memories may ensure we are "stamped with selfhood," but they also may offer an escape. Contemporary brain scientists have much to offer on the connections between memory and the kind of imaginative projection essential for fiction: something that many of today's poets and novelists are acutely aware of.

Modes of Memory

Marcel Proust's epic *A la Recherche du Temps Perdu* (*In Search of Lost Time*, 1913–27) is regularly cited for its penetrating fictional analysis of memory, including its subtle relationship with perception. It also demonstrates the ongoing, dynamic reciprocity between literature and neuroscience. Proust's conceptions of memory were influenced by his engagement with neurological practitioners of the early twentieth century—such as his own father. His *magnum opus* then went on to influence not only other writers, but also neuroscientists.

In his vast novel, Proust anatomized the distinction between two forms of memory: voluntary and involuntary. An act of *voluntary memory* would be, for example, when we reminisce about an episode from our past, perhaps one that brings us joy.[1] By contrast, *involuntary memory* is recall triggered by a stimulus because of an association of which we may not even be consciously aware. The taste/smell of a tea-soaked madeleine in *Swann's Way*, the first book of *Lost Time*, unleashes an unanticipated surge of related recollections and ruminations. Subsequent research has revealed that Proust was pointed toward this distinction through his contact with Dr. Paul Sollier, an associate of the great French neurologist Charcot and an expert on hysteria and memory. Proust even took a "rest cure" with Sollier that involved so-called emotional revivals as part of the program (Bogousslavsky 2007).

One who fell under the sway of the French novelist was the young Samuel Beckett. In his essay "Proust," Beckett described the concept of voluntary memory as "turning the leaves of an album of photographs" as opposed to involuntary memory, which appears as "an unruly magician [that] chooses its own time and place for the performance of its miracle." In Beckett's summation, "The famous episode of the madeleine steeped in tea would justify the assertion that this entire book is a monument to involuntary memory and the epic of its action" (Beckett [1931] 1999: 32). The concept of memories as photographs of the past is widespread, but is rightly seen as severely deficient since the work of Bartlett in the 1930s and even more so given recent advances in neuroscience.

Attilio Favorini (2011) and others have pointed out a recursive structure within Beckett's own treatment of memory: repeating *leitmotifs* within Beckett's work from the 1930s to the 1950s, like the novel *Murphy* and the play *Waiting for Godot*, turn up in other works decades later. Furthermore, Favorini has suggested a close correspondence between Beckett's view of involuntary memory and what neuroscientist Gerald Edelman called the "remembered present." According to Edelman, perception in the moment and the recall of stored perceptual images happen almost simultaneously. Indeed, as we will see, the means of recalling images involves playing stored information back through *perceptual* circuits. Thus, it can be difficult to separate the two events (Edelman and Tononi 2001, and below). To an extent, remembering *is* seeing, smelling, tasting—re-experienced.

A Taxonomy of Memory

In Chapter 4, we saw how the processing of visual and auditory stimuli is not monolithic. There are really multiple visual and auditory "systems." Does the same hold for memory? Language, as so often, provides clues. The Italian language has a tripartite scheme that captures the essentials: "In literary Italian ... memories of times past can be summoned up by three words and in three modes—*rammentare* (with the mind, for facts), *ricordare* (with the heart, for feelings), and *rimembrare* (with the body, for physical sensations)" (Hales 2009: 28). These terms provide a clear trail to approach a current taxonomy of memory; we will concentrate here on memories of the mind, of which we are generally aware and are thus called *declarative* or *explicit*. Given that we have already touched on semantic memory in Chapter 6, here we will mostly discuss episodic memory.

Whence the Engram?

The information to the right in Figure 9.1 shows currently recognized types of memory and indicates important sites associated with them in our brain. This understanding came from many experiments and we will mention only one classic series of studies and then sketch out the current neurological picture. Psychologist Karl Lashley worked from about 1930 to 1950, searching for the *engram* or the physical trace of a memory. However, many studies of rats with various-sized cortical lesions failed to reveal one specific site where memory for navigating a maze could be disrupted, the only general finding being that there were often effects with the largest lesions. He concluded that the memory he was studying was not localized, but rather distributed widely across an "equipotential" cerebral cortex, and assumed that cortical involvement was based on "mass action" where memory cells might be spread diffusely across the cortex. The initial premise behind these experiments, summarized in *In Search of the Engram* (1950), was that memory was a unitary phenomenon focused exclusively on the cerebral cortex.

In line with the taxonomy in Figure 9.1, we now believe this is not the case: memory occurs in several distinctive forms, and it can depend upon both cortical and subcortical structures. Lashley went down a *cul-de-sac* because he chose a behavioral task that is complex, having some dependency on cortical and subcortical structures, and being strongly procedural or implicit. He was only correct in the broadest interpretation of his conclusion: memories are not stored in one place. Yet as Figure 9.1 shows, there are specific sites where key memory functions occur. While the cortex is indeed involved in some forms of memory, it is *not* equipotential.

In a recent success for brain science, cellular and molecular techniques have been used to track down how different memories are apportioned

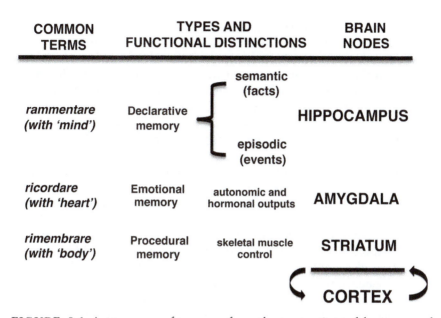

COMMON TERMS	TYPES AND FUNCTIONAL DISTINCTIONS		BRAIN NODES
		semantic (facts)	
rammentare *(with 'mind')*	Declarative memory		**HIPPOCAMPUS**
		episodic (events)	
ricordare *(with 'heart')*	Emotional memory	autonomic and hormonal outputs	**AMYGDALA**
rimembrare *(with 'body')*	Procedural memory	skeletal muscle control	**STRIATUM**
			CORTEX

FIGURE 9.1 *A taxonomy of memory from the perspectives of language and brain science. Linguistically distinguished memory types at left. Categories known from neuroscience are in the middle, with some brain "nodes" of importance to the far right. This is not an exhaustive list, but covers the main forms. For example, emotional memories can have both declarative (explicit) and procedural (implicit) associations. The hippocampus and amygdala are discussed at several points in the text; the striatum is part of the basal ganglia. Cortex here stands for various "association" areas, that is, beyond primary cortex and the arrows indicate two-way interactions between cortex and these nodes.*

within brain circuits, leading to a paper with the playful title, *Finding the Engram* (Josselyn et al. 2015), and a number of other important publications showing how memories can be selectively deleted or altered at the cellular level (e.g., Tonegawa et al. 2015, Kitamura et al. 2017).

Neural Dynamics of Memory

According to modern cognitive views of memory, following sensory and perceptual experiences our brains first keep them, or a filtered record of them, in a buffer—*short-term memory* storage. Here, we hold items at the ready until we can act on them, or alter them for longer-term storage. A phone number is a classic example of a short-term memory item, but it could also be the directions you receive to navigate a hallway toward the fire exit. As one neuroscientist summarized it, "Short-term memory is the immediate present, the information on our radar screens at this very moment; it expires within about 30 seconds or less, depending upon the task. Its capacity is limited, and it fades immediately if we do not rehearse it or convert it into a form that can be retained in long-term memory" (Corkin 2014: 52).

A related, but distinct concept is that of *working memory*. "Working memory is short-term memory on overdrive. Both are temporary, but short-term immediate memory is the ability to reproduce a number of items after a short or no delay (like saying 3-6-9), whereas working memory requires storing small amounts of information and simultaneously working with the information to perform complex tasks (like multiplying 3×6×9 in your head)" (Corkin 2014: 65).

The amount of information that can be successfully maintained in short-term memory is quite limited: the typical form of phone numbers—about 7 digits—was chosen to respect that fact. Prefrontal cortex is known to be a site where short-term memory resides, and it seems to involve the ability of neurons there to maintain a heightened level of electrical signaling activity until influence can be exerted on more posterior parts of the brain. Several other cortical regions also participate in short-term memory storage.

Experiences can eventually be stored in a less volatile form as *long-term memories*. The upper limits on the amount of information that can be retained in long-term memory are unclear, but potentially vast—cases of truly extraordinary memory capacity have been documented. These are known from clinical research and have been the focus of literary fascination as well.[2] Such memory covers a number of distinct cognitive functions, as displayed in Figure 9.1.

Bodily memories (sometimes called muscle memories) are properly categorized as *procedural* or implicit. These can entail the retention of a habit for hours, days, or a lifetime: a person learning to play an instrument or type on a keyboard. Bodily memory also includes simple conditioned associations and usually occurs with little conscious awareness.

Quite distinct from bodily memories are memories "of the mind": *declarative*, or explicit, memories. There are two main types that relate to language and literature: *semantic memory*, our mental dictionary of words and facts, and *autobiographical* memory, our record of experienced events. We will consider these aspects of explicit memory below.

The final mnemonic category is "of the heart," or emotional memory. There are many factors influencing how emotional experiences become encoded and, conversely, how our affective states bear upon memorization. Both aspects of emotion were touched upon in Chapter 7.

Mnemonic Circuits

If Lashley's ideas came from a faulty model, then we must ask how are memories actually distributed and what degree of localization exists? While memories can be widely distributed, such distribution is not random. For instance, there are cooperative interactions between different cortical and subcortical regions, establishing several memory networks (right side of Figure 9.1). In addition, there are a number of intricate cellular processes that act in temporal and spatial coordination to bring about the underlying changes involved in memory formation, sometimes involving the strengthening of specific neural connections and even the creation of new synaptic contacts.

From a functional perspective, the secure cellular changes entailed in long-term memory are a Rubicon: they require the synthesis of new protein in the brain, and alterations in DNA as well. If that fails to occur, a memory remains volatile and will be lost (Kandel et al. 2013, chapter 67, or Heyward and Sweatt 2015).

Figure 9.1 also shows what we mean when we say memories are not randomly distributed: there are distinct subsystems, but with much interplay between them. For example, in the case of episodic memories—including those that are autobiographical—there is a close relationship between the hippocampus, on which new memories depend, and the cortex where consolidated memories mostly live. As Eric Kandel remarked for long-term memory, we have evidence "that [it] is stored in the cerebral cortex ... in the same area ... that processed the information" (Kandel 2006: 129). This is entirely consistent with what we described for semantic memory in Chapter 6.

So, multiple corresponding cortical regions participate: olfactory for odor-related events, occipital cortex for visual events, and so on. One area that has an especially significant role in most aspects of declarative memory function is an extended medial temporal lobe (MTL) system—of which the *hippocampus* is a part. The two other forms of long-term memory involve significant subcortical circuitry: procedural memory—the striatum—and emotional memory—the amygdala. You can see the approximate location of the hippocampus and amygdala in Figure 7.2.

As opposed to the largely unconscious nature of procedural (habit-based) memories, we have conscious access to episodic and semantic ones. However, they still have pre-conscious elements. As we all know, it is possible to store information about facts or events but be unaware of them until something triggers their recall—that esoteric crossword clue—only then do they rise to awareness. The mysteries of conscious awareness and recall were succinctly framed by Byatt (2008: xiii): "searching for names 'I know I know' (and *how* do we know we know if we can't remember?)."

By asking people to make so-called remember-know judgments, neuropsychologists have helped solve this puzzle: it turns out remembrance for words or events can come in two basic types. Consider a simple case where people learn a list of words. You can mix these randomly with a new list and then ask them if they recognize words as being new, or from the memorized set. If a word is identified as one previously learned, you then ask for a second decision: is the word merely known ("I recognize it") or is its acquisition actually recalled ("I remember reading/memorizing it")?

The first is something like the situation Byatt described, a tenuous sense of familiarity; the second is the true recall of episodic memory. People can easily distinguish between these two—and the brain can be scanned functionally as they do so.

Early imaging studies with fMRI indicated that the hippocampus is strongly activated during episodic memory recall, but not when a person only experiences a bare sense of familiarity (Eldridge et al. 2000, Buckner 2000). So some have suggested that the hippocampus mediates episodic recall, while a nearby region ("perirhinal" cortex) is responsible for mere familiarity. Intriguingly, studies of epileptic patients who report *déjà vu* episodes prior to seizures indicate a link between these and our sense of familiarity (Martin et al. 2012). Yet the relationship of familiarity to recall in the medial temporal lobe-hippocampal region is hard to disentangle—this is a story still under refinement.

A growing consensus from neuropsychology is that the MTL region around the hippocampus is activated *both* during sensory experiences and also when we recall those experiences (e.g., Zeidman and Maguire 2016). These striking findings provide confirmation of something long suspected by those engaged in understanding memory, "memographers," both literary and scientific: that when we remember an episode we are in effect re-experiencing (or "replaying") a version of it—hence the many pleasures and pains of our mental life. As Andrew dryly observes in Doctorow's novel, speaking of the sudden death of his second wife, Briony, "thoughts of Briony gave me all sorts of perceptive advantages. It was as if something of her mind was still alive in me. Is this cognitive science? Not really. It's more like suffering" (180).

Concepts in the Brain

Discoveries about the overlap between the functions of sensation and recall are an important breakthrough, as far as it goes, but can it be followed up at a more detailed neural level? Well, neurosurgeon Itzhak Fried, along with a group of colleagues, has been doing just that. Prior to performing surgery for epilepsy, they made recordings from electrodes implanted directly in patient's brains to improve surgical accuracy. Fortunately, these electrodes also allowed the researchers to investigate whether perceptual processes really do overlap with memory recall (Gelbard-Sagiv et al. 2008, Suthana and Fried 2012).

The study participants viewed a randomized sequence of brief cinematic clips as the research team recorded neurons in the hippocampus and surrounding MTL. In total they recorded about 850 individual neurons from the brains of thirteen patients. They found neurons that increased their "spiking rate" and maintained the increase for an extended period only while *certain* video clips were being watched. Then after a rest, patients were encouraged to talk about what they remembered. Neurons that initially "spiked" on viewing a clip began firing intensely again—but only when that particular clip was recalled.

One patient had a cell that responded specifically to a clip from *The Simpsons*. When freely recalling *The Simpsons* episode, but not other clips, that particular cell issued a burst of spikes once more. The timing of the neuron's firing made sense—it increased a few seconds before the patient described the recall (Figure 9.2). This was not an isolated occurrence: more than half of the total set of 850 neurons they recorded showed such selectivity to sensory responses and were re-activated during recall.

This episode-specific reactivation was found in the hippocampus and adjacent MTL. There are several remarkable features about this experiment. First, there was no cue provided by the experimenters—episodes were recalled and described spontaneously. As a result, if you were monitoring

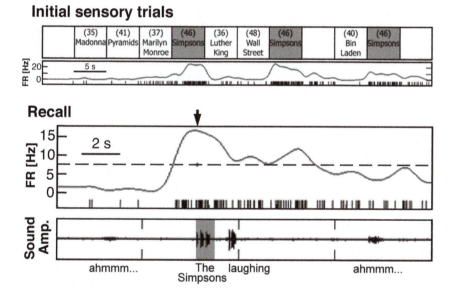

FIGURE 9.2 *Neurons of recall and knowledge.* **Upper panel:** *Trial-by-trial response of a neuron recorded from right medial temporal cortex. Order of clips (at top) is for the purpose of illustration. Instantaneous firing rate [FR in Hertz] is displayed at bottom along with the record of neuron impulses—short vertical lines. Note that only a few impulses occurred during most segments, but a large number reliably occurred during each Simpson's clip.* **Lower panel:** *A recall event shown at greater time resolution. Sound bursts from a voice recording are shown (bottom trace)—the content of the voice recording given at very bottom. Note that patient says it's* The Simpsons *he is thinking about just after the neuron increases firing. Time calibration bar = 2 sec. From: Gelbard-Sagiv et al. (2008) Internally-generated reactivation of single neurons in human hippocampus during free recall.* Science 322:96–101, *adapted from Figure 1B and D. Used with permission.*

the impulse activity of the set of cells in a patient, you could predict when that patient was recalling an episode, and even which one, even before they spoke!

Given that a number of video clips were used to probe each neuron (typically forty to fifty), and that only a few activated neurons strongly, the experimenters commented:

> "The neuronal responses reported here were extremely *sparse* and seen selectively in individual neurons out of dozens of nonresponsive neurons that were recorded in their immediate vicinity." So it was clear, "By a simple statistical reasoning, that these (nonresponsive) neurons must display selective responses to multiple other clips that have not been presented, as only a minute fraction of the vast set of possible episodes was tested in our study." (Gelbard-Sagiv et al. 2008: 100)

The sparse coding described here, where a few cells among many can encode a stimulus, is similar to what was observed in face-selective areas in the ventral visual stream (Chapter 4). There is no evidence yet that any patient had but a single "Bart Simpson neuron," or "Jennifer Aniston neuron"—as would be the case if the extreme "grandmother cell" model applied. Rather there seems to be a relatively small group of cells that encodes any one event or character, distributed among many others with different selectivity. The coordinated activity of such a group of neurons would then constitute the experience, or re-experience, of an episode or a specific character. This dramatic finding posed another question: how do cells in a region like this become tuned to a precise stimulus in the first place?

To address this, the team extended their findings to observe neuronal selectivity in the process of forming, as people encountered new stimuli. Like before, they recorded from MTL neurons. Patients were screened during initial electrophysiological recording using a large number of visual images to find out what specific stimuli neurons responded to. Subsequently, pictures were selected (of a person for whom a neuron had been selective) and it was then paired, in a composite image, with another picture (of a location) to which there had been no response. As an example, a patient with a neuron selective for a picture of actor Clint Eastwood was shown a picture of Eastwood and the Eiffel tower, a location that cell had not responded to. Revealingly, neurons gained selectivity for these new combinations with the same speed (a few minutes) as patients gave behavioral evidence of the new association (Ison et al. 2015).

In a further intriguing development, this team has also uncovered how the responses of neurons in the MTL region may "generalize." For example, neurons responsive to a person—a picture of Oprah Winfrey—responded to any of several pictures of her: in different dress, different environments. This is what a key author on the study, Rodrigo Quian Quiroga, calls "visual invariance" (Quian Quiroga et al. 2009). The phenomenon suggests at least one possible answer to the ancient question in epistemology: how can we have general knowledge of the world? How do we recognize a person, animal, or object when their appearance is never really identical any time we see them? Many have speculated that our mental processes for "object recognition" must entail generalization. These MTL neurons take the discussion beyond mere speculation.

Visual invariance helps explain generalization within visual perception *and* memory. What Quian Quiroga also observed was that about one third of cells responding to pictures of people also responded when viewing text, and another third to sound. About 20 percent responded to all three modalities congruently. So in one example, they found a cell that responded to pictures of Oprah Winfrey, also to the name "Oprah" as text, and to acoustic presentation of a voice saying Oprah's name. Such a cell then can signal "Oprah" through any of three perceptual channels. These cells go beyond mere visual coding and show *multimodal invariance*. In such a sparse coding scheme, it is likely that a small assembly of neurons encoding, say, one particular person, will contain neurons responding to a different constellation of stimuli related to that person. This is the basis on which your "concept" of them transcends one sensory modality. Quian Quiroga has called such cells "concept cells" (2012b).

Some of the neurons in these studies responded specifically to family members or friends, but many were triggered by famous people that the patients would not have met: Vladimir Putin, Salma Hayek, or locations that patients had not visited, and some to fictional characters. These results offer a rich set of possibilities for speculating about the associations we form in memory, and how perceptual and rememorative events may underlie reading.

The Literary Moment

The findings summarized above are the beginnings of an explanation for what I. A. Richards hypothesized: that there must be some traceable way in which lines of text influence memory and emotion—and, conversely, by which these modify our response to text. Of course, some of these are easy to observe in ways that do not require access to brain imaging devices: ask readers.

Keith Oatley described a simple technique he used: give readers a story and ask them to write a mark in the margins any time they experienced a flash of memory or a pulse of emotion. Before explicitly discussing either memory or emotion in literature, people were asked to read a short story and to provide their reactions. They displayed both types of responses, triggered at many points throughout the story (Kate Chopin's *Dream of an Hour*). This simple approach demonstrates that memories and emotions are evoked in reading and may help to isolate some of the linguistic triggers.[3]

Quian Quiroga's experiments tested MTL neurons with a large number of stimuli—using these data to then speculate on how networks of "concept cells" might be activated to represent characters in stories. One person had cells that responded to images of Luke Skywalker and Yoda from *Star Wars*. So Quian Quiroga constructed a model of how neuron clusters with these two specificities might overlap with each other, and with neuron clusters for other characters from the *Star Wars* saga.

There is behavioral and brain scan evidence suggesting that we respond in similar ways to characters created through different media. If our neuronal reactions to characters (and events) in books are similar to those for characters and events presented cinematically, then his model provides at least one neuron-level model for understanding our "transmedia" reactions to story elements in fiction.

So if one probed the brains of experienced readers, perhaps the names Emma Bovary, Anna Karenina, or Stephen Daedalus (or images of them) might elicit activations of MTL or similar neurons. In Figure 9.3, we show a simplified schematic of how one might think about using Quian Quiroga's approach. The figure shows three cell assemblies—each one containing neurons responding primarily to one important character from *Ulysses* by James Joyce. A hypothetical neuron assembly for Leopold Bloom is shown centrally and it overlaps others for his wife, Molly Bloom, and his companion for a day, Stephen Daedalus.

Given that concept cells for specific individuals can be engaged in a multimodal manner, the activity in such a network may ultimately arise from visual images of people of a certain physical type as fictionally described—with gaps for us to fill, perhaps from people we know, the book cover, a film version, or auditory or textual associations.

Experiments of this sort could be done to probe the readerly "mind's eye," or "mind's ear." A word of caution: we are considering here what might be going on to bring vivid characters to life in our minds. Such speculation should not be taken as suggesting that such a neural network would "explain," let alone "explain away," the literary and emotional impact of *Ulysses*. What we are suggesting is that plausible models are now available for at least some aspects of our "storyworlds" at the neuronal level. To take it a step further, since it has been shown that neurons in MTL can be changed by new stimuli within the narratively relevant time frame of minutes, it is natural to wonder if such modifications might happen during a single bout of reading.

Or, because we know readers approach any work with pre-existing memories (i.e., neuron specificities) for familiar individuals—might this be why fictional characters frequently resemble people we know? Are our character-images neuronal "echoes" of our concepts of real people preserved in networks of this sort?

More broadly, one could ask, too: how would such network models relate to the literary speculations described in our opening chapters? Is the deep

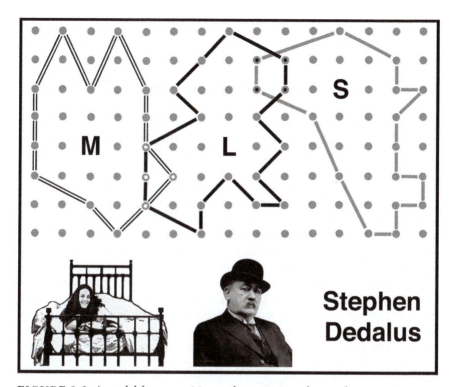

FIGURE 9.3 *A model for recognition and association of story characters. Diagram shows how neuronal assemblies of "concept cells" might be applied to an aspect of literature. Three hypothetical cell assemblies are shown encoding "concepts" from* Ulysses. *Each gray dot stands for a neuron and each set interconnected by specific types of lines represents one assembly. The panels below suggest the multimodal nature of the assemblies: neurons in the Leopold Bloom assembly (L) fire for a visual image of that character, the Molly Bloom assembly (M) has cells that fire to an important locational cue (the bedroom), Stephen Dedalus (S) assembly cells are selective for the text of the character's name on the page. Notice there is some overlap between the concepts (assemblies): so a few neurons fire for cues related to either concept, and might encode associations between concepts. Inspired by the data/model in Quian Quiroga 2012b: 2016.*

perceptual structure posited by Elaine Scarry analogous to what's described here by Fried and Quian Quiroga?

Consider that memory *mimics* perception. Remembering an event (sunlight glinting on seawater, the taste and feel of cold beer on a hot day) is physiologically *almost* the same as re-experiencing it, but with less immediate intensity. Certainly, memories are triggered by reading, and, as seen, the deep perceptual processes engaged can be choreographed by an authorial "score," which orchestrates specific mnemonic/perceptual events across reading time. If this is the case, then "imagining" would be a form of perceptual mimesis, whether in a dream or under the instruction of skilled writers. Neuroscience is not so very far from testing such speculations.

We Don't Know What We Don't Know

Neuroscience has managed to catch up with questions about why memory is inexact, protean. We now have strong evidence that recall is, by its very nature, constructive and even re-constructive. This was first articulated scientifically by Frederic C. Bartlett in his classic *Remembering* ([1932] 1995), a summary of experimental findings. Importantly, his work not only sheds light on the characteristic ways we evoke the past but also suggests why our memory processes may be related to thinking about the future.

Bartlett wished to investigate memory under everyday conditions, so focused on how people remember stories. He studied how accurately they could be recalled after various lengths of time, and how recollections changed when repeatedly brought to mind. His summary: (1) memory is fallible— "accuracy of reproduction in a literal sense is the rare exception, not the rule"—(2) recall is incomplete, prone to "omission of detail, simplification of events and structure, and transformation of items into more familiar detail," and (3) at some point we begin to invent. In remembering the distant past, "elaboration becomes rather more common," with increasing "importation" and "invention" ([1932] 1995: 93).

If this sounds unsettling, it need not be. Bartlett put it all in a broad perspective: "In a world of constantly changing environment, literal recall is extraordinarily unimportant" (204). Furthermore, the fact that memory is constructive enables us to appreciate the relation of memory to imagination which is also constructive.

A fine way to study mnemonic unreliability is to use so-called flashbulb memories. These are vivid personal recollections of surprising public happenings ("when did you first hear about ...?"). A recent study analyzed people's memories for the September 11 attack in the United States. They reported, "There was rapid forgetting of both flashbulb and general event memories within the first year, but the forgetting curves leveled off after that, not significantly changing even after a 10-year delay. Despite the initial rapid forgetting, confidence remained high throughout the 10-year period" (Hirst et al. 2015: 605). We tend to be more certain of our memories than we should be.

In this regard, psychologist Ulric Neisser, often credited with founding cognitive psychology, told a wonderful anecdote:

> I long remembered how I had heard the news of the Japanese attack on Pearl Harbor on December 7, 1941—the day before my thirteenth birthday. I was listening to a baseball game on the radio ... The announcer

interrupted the game to report the attack … A perfectly good flashbulb memory, one would think. I thought so myself for twenty or thirty years, until one day it occurred to me that no one plays baseball in December! (Neisser 2008: 82)

He was later told that there was a big football game that day, so he probably confused his sports, not the entire memory episode. To make sense of errors in memory, and also the logic of imagination being "memory cast forward," it helps to put the dynamics of memory in its behavioral/physiological framework.

The Gist of the Story

What might explain such disappointing memory performance? The reasons lie in two basic facts: our perception is constrained by our Umwelt and our attentional resources can't be focused everywhere at once, so we necessarily retain a partial record of events—the "gist" of what happened. This record can be stored in a stable ("consolidated") form over long periods of time. Then at recall, the memory becomes fluid again as we reactivate and re-experience elements of it. It has to be stored again ("reconsolidated"), and sometimes its accuracy is degraded further, or it can even have new elements (aspects of false memory) added to it. In some respects, the process acts like a bad photocopier, churning out increasingly distorted images. As a result, the metaphors (a filing cabinet, camera, audio recorder) that have been applied to memory all fail to capture its essence. A more accurate metaphor would be memory as narrative, as a message in the game telephone: half-reproduced, half-created.

Bartlett's observation that "literal recall is extraordinarily unimportant" prompts us to ask about the inherent value of "creative" recall. One take on the implications was suggested by Daniel Schacter and Donna Rose Addis: "Many researchers believe that remembering the gist of what happened is an economical way of storing the most important aspects of our experiences without cluttering memory with trivial details." They also note that we rely on past experiences to predict the future—but:

Future events are not exact replicas of past events, and a memory system that simply stored rote records would not be well-suited to simulating future events. A system built according to constructive principles may be a better tool for the job: it can draw on the elements and gist of the past, and extract, recombine, and reassemble them into imaginary events that never occurred in that exact form. Such a system will occasionally produce memory errors, but it also provides considerable flexibility. (Schacter and Addis 2007: 27)

The Remembered Future

We can now begin to relate constructive memory to larger cognitive issues. In Eric Kandel's summary of research on memory, he pointed out its transcendent quality: "Remembering the past is a form of mental time travel; it frees us from the constraints of time and space and allows us to move freely along different dimensions" (Kandel 2006: 3). Both Doctorow's Andrew and McEwan's Henry Perowne spend much of their fictional lives engaged in just such time travel, as do we. One of these "different dimensions" of mental journeying is *prospection* rather than *retrospection*—the ability to cast memory forward (overlapping with imagination but not identical, see below), to envisage the future, or even something dreamed, or perhaps not even possible. Of course, there is a long-standing association of memory with imagination. Aristotle, among others, believed, "It is obvious … that memory belongs to that part of the soul to which imagination belongs" (Harvey Wood and Byatt 2008: 159), while Vico advised, "The teacher should give the greatest care to the cultivation of the pupil's memory, which, though not exactly the same as imagination, is almost identical with it" ([1709] 1990: 14).

Surprising support for this view has come from two areas. First, when researchers took seriously a possible overlap of neural substrates for memory and imagination, it was discovered that damage to the hippocampus, known to produce amnesia, also impairs the ability to imagine personally relevant future events, or describe fictitious scenes (Hassabis et al. 2007).

Second, with functional imaging it was possible to locate brain regions active during recall and compare them directly to a map of regions activated during imaginative projection into the future. Karl Szpunar and colleagues, and, separately, Donna Rose Addis and colleagues, reported that several areas of the brain displayed activity both when someone imagined future personal events and when they remembered past ones. These regions were the left hippocampus, posterior cingulate cortex bilaterally, parts of right parietal cortex, select parts of the occipital cortex on both sides, the precuneus on the left side, and medial prefrontal cortex, see Figure 9.4 (Szpunar et al. 2007, Addis et al. 2007). The areas activated in each case were not identical, but highly overlapping.

Consequently, there has been a considerable upsurge of work seeking to understand future-directed cognition. What is crucial is to distinguish varied types of such "prospection," a term which can cover goal-setting, planning, prediction, and simulation; some of them are highly autobiographical while other aspects are more abstract (semantic). If, as I. A. Richards once commented, "a book is a machine to think with" ([1924] 2001: vii), then the ability of narratives to stimulate memories (both autobiographical and semantic, prospective and retrospective) may be at the very core of their imaginative capacity to open up time (and space) travel.

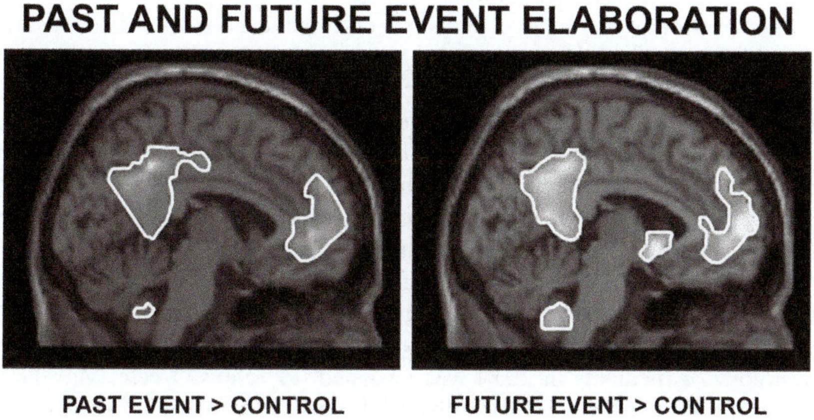

PAST AND FUTURE EVENT ELABORATION

PAST EVENT > CONTROL   FUTURE EVENT > CONTROL

FIGURE 9.4 *Time's Arrow Visualized. MRI activation patterns for recalling past events compared with imagining future events. Midline slices illustrating the striking commonalities in medial left prefrontal and parietal activity during the elaboration of (a) past and (b) future events (relative to the control tasks). From: Addis et al. (2007) Remembering the past and imagining the future: Common and distinct neural substrates during event construction and elaboration.* Neuropsychologia *45:1363–77, Figure 2. Labels enhanced and area of activity in color original outlined for clarity in B/W version, used with permission.*

Integrating Temporal and Spatial Domains

Before we can really understand how a prospective dimension fits into memory function, we need to develop some of the background on the center in the brain that figures large in the comprehensive story of memory. The reputation of the hippocampus rests on a strong experimental base relating it to memory. Most tragically, it relates to two individuals with significant hippocampal damage: Henry Molaison in the United States who underwent bilateral hippocampal removed in the early days of epilepsy surgery and Clive Wearing in the UK who sustained hippocampal damage from encephalitis. Both men lost the ability to form long-term episodic memories and so lived in an eternal present, with perhaps 30 sec of short-term recall. Yet they both retained procedural memory, showing that these different forms of memory are neurologically distinct. As well as technical reports on these men, there is also a novel inspired by their cases (*The Man without a Shadow*, Joyce Carol Oates [2016] 2017).

Other evidence of a hippocampal role in memory is far from tragic, and even astounding. As mentioned in Chapters 2 and 7, the "brain space" for cognitive functions is subject to variation within one's lifetime and the hippocampus has provided several good examples. One is related to reading and memory: professional pandits in India who spend years memorizing tens of thousands of pages of Sanskrit texts have been found to possess larger hippocampi, and also larger language areas in the temporal lobe than controls matched for age and number of languages spoken (Hartzell et al. 2016). Also in a now-famous example, the hippocampus was found to be larger in London taxi drivers who had mastered the complex street map of the city (e.g., Maguire et al. 2003). This is believed related to the repeated memory demands of this job, but knowing a city, and moving passengers through it, requires another skill: spatial navigation.

Additional evidence, recordings from neurons, have indicated that the hippocampus—in addition to guiding memory formation—is also fundamentally involved in spatial navigation. In fact, the Nobel Prize for physiology or medicine in 2014 was awarded (to John O'Keefe, May-Britt Moser, and Edward Moser) for influential work on how hippocampal cells encode position in space. These cells were found in rodents, but there is evidence that humans have similar but even more sophisticated hippocampal systems (e.g., Rolls and Wirth 2018).

For a while there was a lot of speculation about how these different behavioral processes could be understood as products of only one brain area (e.g., Buzsaki and Moser 2013). Yet they undoubtedly are, and this linkage is beginning to provide hints on ways to think about spatial and temporal aspects of cognition generally and (so relevant here) stories in particular.

To start, the hippocampus is actually part of one subsystem within the default mode network (DMN, Chapter 6). So the hippocampus and associated structures are selectively active during self-referential thought and construction of visual and multisensory scenes, which explains why hippocampal damage degrades the ability to recall spatial scenes or construct imagined ones.

These findings caused Eleanor Maguire to propose the "scene selection theory" of hippocampal function (Figure 9.5, reviewed in Mullally and Maguire 2014). The idea is that the function of this region is not to represent immediate sensory space itself, but to *direct* the construction of scenes or episodes from information stored elsewhere. Such scenes are indispensable for remembering the past, *and* imagining the future, as when planning a trip or navigating on one. Michael Corballis summarized the situation this way:

As animals that move, we are supremely adapted to knowing where we are in space, as well as knowing where we have been and where we might go next ... we can travel mentally in space and time, replaying past episodes in our lives and imagining future ones, bringing to mind events that have not actually occurred. This is important not only for recording which places and events are dangerous or fruitful, but also for planning futures, perhaps mentally creating different possible scenarios to establish choices for future consideration. Mental travel can also free us from reality, *allowing us to create stories and imaginary adventures*. (Corballis 2018: 589, our italics)

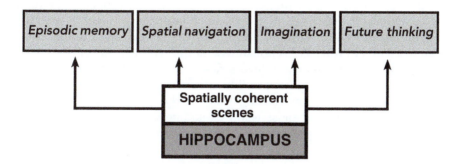

FIGURE 9.5 *Multiple functions of the hippocampus are integrated in scene selection theory. This has been formulated by neuroscientist Eleanor Maguire and colleagues. From: Mullally and Maguire (2014) Memory, imagination and predicting the future: A common brain mechanism?* The Neuroscientist 20:220–34, *Figure 6, lines enhanced for clarity and hippocampus label added. Available through Creative Commons attribution license 3.0.*

The recent synthesis of ideas about what the hippocampus may be doing as part of the DMN has echoes of the ideas of literary theorist Nancy Easterlin (2012). She suggested that our nature as a wayfinding species is connected to our tendency for spatial thinking and planning—and for storytelling. Current neuroscience has provided a large amount of data consistent with this view and accorded it a potential hot spot at the nexus of the MTL system and the DMN network of the human brain.

The Narrated Self

The very terms used for our personal memories—episodic and autobiographical—suggest that what is remembered has a narrative quality. Indeed, those who have thought about memory and personal identity have often assumed that the features of narrative structure can open a way into the workings of memory, and, thereby provide an integrated understanding of imagination and the sense of self. Much of this work has been done by psychologists, such as Jerome Bruner, but also by philosophers, such as Marya Schechtman and Peter Goldie, keenly aware of the findings emerging from cognitive neuroscience. Their work illustrates how the mechanics of memory are embedded within the larger structures of human cognition, including emotion, and how the study of literature repays careful attention with respect to these cognitive processes.

Schechtman has articulated a position she calls a "narrative self-constitution view" of human identity: "A person creates his identity by creating an autobiographical narrative—a story of his life." By using the term "narrative," Schechtman is suggesting that the persistence of our psychological identity over time depends upon coherent memory facilitated by an underlying linear narrative. Not all philosophers of identity, even if recognizing the value of narrative thinking, agree—seeing a linear framework as restrictive. However, Schechtman asserts that episodes of memory are characterized by different narrative styles and "all of the stretching and redefining of narrative that is possible in literature ... might occur in an individual's self conception," adding "some of the most brilliant moments in fiction are achieved by those who expand our perception of what kind of thing can be a comprehensible story" (Schechtman [1996] 2007: 95, 102, 105, respectively). Here is a vision of the self as a narrative creation, and one moreover subject to the unconscious intervention of "stylistic" techniques—of focalization, point of view, and even modes of speech and thought presentation.

She goes further to argue that the continuity of our present consciousness with our past selves is influenced by (a) the affective reactions we have to memories of our past (which occur in the present) and (b) the fact that our understanding of past events leads to a "script" superimposing coherence on the self through time.

Her inquiry into mind and personhood reinforces the value of drawing concepts from literature. As she put it: "The sense of one's life unfolding according to the logic of a narrative is not just an idea we have, it is an organizing principle of our lives. It is the lens through which we filter our experience and plan our actions, not a way we think about ourselves in reflective hours" (Schechtman 2007: 113).

Peter Goldie differs on some points, nonetheless, he also finds narrative to be a central integrating principle, and holds that at the very least we understand past and future "in narrative terms." Like Schechtman, he believes emotion is integral to grasping how narratives function to support a concept of self. Without such narrative evaluation and emotion, he suggests, our mental experience of the past would be reduced to annals, dry litanies of events. Positing narrative as such an organizing principle of cognition entails looking with fresh eyes at the structural and affective features of literature.

Goldie provides examples to show that our stories of the past may be voiced, but may also remain private without violating their narrative nature. In this way, he clearly lays out a case for *narrative thought* as fundamental to the self. In doing so, he recognizes the heterogeneous content of memories, from those that are deeply experiential to those that are merely semantic (we may vividly remember our celebration of a friend's birthday, or merely recall *that* we celebrated her birthday). They may include imagistic slices like flashbulb memories, event-specific recall in classic "episodic" form, general events (I was at a party), or the broad recollection of life periods (that era when I partied a lot). All this is woven together as the threads of one person's past via narrative. Finally, Goldie concludes, the entire phenomenon—the panoply of the past—can be "played forward" via the imagination. We can only conceive of the future by deploying past experience, with emotions as our guide.

One of Goldie's distinctive contributions is recognizing the occurrence of ironic distance within memory and narrative thought. This arises, he believes, because we can engage *multiple perspectives* in narrative acts of memory. So for instance, "Someone who is internal to a narrative, having a role as a 'character' ... can also be external to it, having also the role of external narrator. This is the case in autobiographical narratives" (Goldie 2012: 26). But this is also true (perhaps especially true) of the "hall of mirrors" that is fiction:

> Novels are the most wonderful source of the phenomenon: in novels there can be diverging perspectives between two or more characters ... between character and external narrator, between external narrator and internal narrator, between character and reader ... between external narrator and implied author, between implied author and author, between implied author and reader, between reader and reader, between actual reader and ideal reader, and many more besides. (30)

We now see why the narrative "slipperiness" of both *Saturday* and *Andrew's Brain* is more than authorial play, but derives from the fact that they are both, profoundly, novels about the sense of self, about how our brain constructs it.

Consider that as the person recalling a memory in the present, you will have knowledge or emotions that the character (perhaps you) did not have during the original experience. This is how dramatic irony arises in plays and novels where the reader and characters have incompatible beliefs or knowledge. As Goldie says, "There is, in autobiographical memory, an ineluctable ironic gap (epistemic, evaluative, and emotional) between internal and external perspective, and ... our memories are themselves infused with this irony through the memory equivalent—the psychological correlate—of free indirect style" (Goldie 2012: 43). As we've seen, this stylistic device entails just such a melding of first- and third-person perspectives. There is now empirical support from psychology to suggest that visual components of autobiographical episodes are retrieved from either a first- or third-person perspective, and that any given episode may show a mixture of the two (Rice and Rubin 2009, 2011).

Schechtman and Goldie have built a sophisticated approach to delineating how different components of memory are related, the ways in which emotion matters, and how literature casts a powerful light on these aspects of the self.

However, not all philosophers agree. Galen Strawson (2015) for one has stated plainly on the basis of introspection, "I am not a story." Even if he is entirely right about himself, it doesn't necessarily undercut the basic stance of Schechtman and Goldie. Here, neurological individuality and personal variability must be considered. People range in their autobiographical memory on a scale from highly superior to severely deficient (LePort et al. 2012, Palombo et al. 2015), raising questions about how individuals with minimal ability in this respect would construct and maintain a sense of self. One relevant perspective here is that recollected information may move between memory categories; for instance, some initially autobiographical events seem to become "semanticized" and thus recalled through means less personal than autobiographical memory (e.g., Moscovitch et al. 2016, Morissey et al. 2017).

Of all the implications of recent findings on memory and imagination, the most interesting—besides some obvious ones in mental health—would have to be the implications for understanding our storied minds. We can now see that imagination, like memory, comes in forms both implicit (such as rapid construction of situation models in reading) and explicit (creating a scene of somewhere you intend to travel, physically or mentally). And while visual imagery has led the way in past research on imagination, future studies will extend beyond visual imagery to multisensory or even synesthetic possibilities. Whatever the outcomes, the work of cognitive neuroscience should continue to expand what we can know about the imagination, and narrative as an exemplar of cognition and the construction of a temporally extended self will retain high value on the "mattering map" of the human sciences.

Notes

1   Proust sought medical help for—among other things—impairment in his ability to electively recall events, at a time when he feared he was losing his creative powers.

2   A classic account is Luria's *Mind of a Mnemonist* (1987). Extraordinary memory related to autistic savant syndrome (and the story of the individual who inspired the movie *Rain Man*) is described in Treffert (2009). For an imaginative literary account, see Borges (1964); and the monograph about Borges by Quian Quiroga (2012a).

3   We tested this approach, using the same segment (eleven paragraphs) of the Chopin story that Oatley used (see Oatley 2011: 64). In a typical case, fourteen students recorded ninety-eight identifiable reactions, split between memories and emotions.

10

The Via Dolorosa of the Self

For our final working study on literature, we turn to a 2014 novel that brings together many of the personal and social implications behind the topics addressed in this book. E. L. Doctorow's *Andrew's Brain* weaves together the strands of unreliable autobiography, our inexhaustible drive for personal coherence, the persistence of the past, the mysterious nature of deep emotions (grief, love), agency and determinism, the mind/brain divide, and even engages with trans- and post-humanism, topping it all off with an apocalyptic vision of the "end of narrative." As if that wasn't enough, Doctorow also has much to say about the risks of a purely reductionist view of life and of ourselves, and the dangers posed to human culture if we do not find a way to integrate our story-driven past with a materialist view of the future.

This work, about a very disturbed cognitive scientist, bears a carefully qualified title. *Andrew's Brain* is not about the lead character *per se*, a fact that becomes increasingly evident as the book unfolds. In the course of it, Andrew even speculates about whether he exists at all, and if so whether it is as the expression of his own DNA, a fragment of collective consciousness, a by-product of the "soundless voices" in his head, or even an artificial intelligence. As with Henry Perowne, the "real" Andrew remains stubbornly obscure, even, or especially, to Andrew himself.

As a title, therefore, *Andrew's Brain* is at once a challenge and a deliberate statement of intent. Doctorow here has created a protagonist whose experience is so calamitous and self-alienating that it undermines every certainty. In a way, the character's raison d'être is as an ontological crash test dummy, a kind of radical stress test for far-reaching ideas of personhood in the age of neuroscience.

This fact also helps to explain the book's structure (or lack of it). In order to delve into Andrew's dilemma, Doctorow uses the device of an unnamed interlocutor, perhaps a therapist or psychiatrist of some kind, to whom Andrew spills out his woes. This framing premise allows a freewheeling conversational movement of the narrative across time—with multiple flashbacks, digressive discussions of cognitive ideas, and airing of personal woes.

As Doctorow revealed, there was another source, derived from his own "autobiographical memories" for this loose, spiraling, structure. "Do you know the music of Steve Reich or Philip Glass? It's phase music in which the music doesn't proceed in a linear organization but as chords or musical statements that slowly, gradually change. Very often music has been an inspiration to me" (interview 2014c).[1] Hence the narrative of Andrew's life does not so much progress as dance backward and forward toward its close in a series of compulsive riffs or refrains. Doctorow wants his novel to mirror the mind's tendency to revisit (and embellish) autobiographical flashpoints. So painful are the events of Andrew's life that he does not have to look far. Indeed, like the tale of Scheherazade, there is a recursive quality to these tales within tales—but on a much more baleful note. For Andrew, misery is heaped on misery to an extraordinary (almost operatic) degree.

In typically playful fashion, Doctorow actually introduces opera into Andrew's life in the orotund shape of his ex-wife's new husband, a baritone. After all, "opera [is] the art of unconstrained emotions" (16a), something Andrew believes he no longer has access to, making Martha's "large husband" a kind of living reproach: "I have no feelings ... I am left cold, impenetrable to remorse, to grief, to happiness, though I can pretend well enough even to the point of fooling myself" (17). The large opera singer is as full as Andrew is empty. As if to compound the point, Andrew's ex, Martha, a concert pianist, is now "irremediably damaged" by the death of her daughter with Andrew, for which he feels crushing responsibility (15).

In these musical allusions, it is as if Doctorow consciously wants to weave his own autobiographical influences into the fabric of his work, a work that, if it was a piece of music, would be a strangely capricious "variation on a theme," set in a minor key. And music, as we've seen, may reach back beyond attributable feeling into "core affect" at a deep, if indefinable, level.

He is unabashed about this process of self-mining, as if constantly goading the reader with the irrational and serendipitous nature of the act of creation itself. *Andrew's Brain* is designed to let us know that it is also about Doctorow's brain, and the essentially mysterious act of its own generation. As we've seen so often, and in many different levels and contexts, introspection can only take you so far.

In keeping with this theme, one of the abiding preoccupations here is with the very nature of thought (those "soundless voices" in our head), which allows the point of view of the narrating voice to mutate and shift, often abruptly and apparently at random from an egocentric to an exocentric mapping of events. As with Henry Perowne in *Saturday*, we are never allowed to comfortably settle into a first-person account of Andrew's life, but rather drift between first, second, and third person, as if to emphasize just how dislocated and dispersed his identity has become, something only accentuated by the passage of time, and the awfulness (sometimes comic, often tragic) of his life story. This is Peter Goldie's "ineluctable ironic gap" between internal and external perspectives in memory writ large, the now-Andrew writhing in discomfort as he relives and recounts the antics of the then-Andrew.

The book occasionally allows us to hear Andrew describing his own feelings, taking ownership of the events that befall him; at other moments, he recounts them from afar, as if the "Andrew" he describes is some other person entirely. At this point, he becomes not just an unreliable narrator in the postmodern style, or one blind to his own narrativity (like Henry Perowne) but one who is pathologically dissociated. The novel begins with Andrew disowning not only his own "proprietary" consciousness, but even his own identity: "I can tell you about my friend Andrew, the cognitive scientist. But it's not pretty" (5).

He then unfolds the tale of someone upon whom every kind of misfortune seems to have been visited, taking the opportunity to goad his therapist along the way:

> I read about the pathogenesis of schizophrenia and bipolar disease. The brain biologists are going to get to that with their gene sequencing ... So, doc, you're in trouble with your talking cure. Don't be too sure. Trust me, you'll be on unemployment. (6)

Already, in these opening pages, the battle lines are drawn, embodied in the wretched case of Andrew himself and the issue of whether reductionism—treating the mind as brain and nothing more—can ultimately solve all our ills in the aftermath of a *deus absconditus*: "God works through Darwin" (6).

Ironically, as Andrew continues to weave the story of "his friend, Andrew" and the terrible events that befell him, it is the granular detail of his descriptions that betrays to the therapist the truth of what is going on. "Are you in fact the man you call your friend Andrew, the cognitive scientist who brought an infant child to the home of his ex-wife? Yes" (9). The implicit analogy between Andrew, who accidentally kills one daughter and gives the second away, and a God who has abandoned his "children" need barely be spelled out, and continues the elegiac tradition McEwan accessed through Matthew Arnold.

Here Andrew, *qua* storyteller, gives himself away by including too much "telling detail," too great a degree of specificity, in the anecdote of "his friend" for the account to really be that of a third person. "It was some sort of a broken-down farm. ... There was an old flatbed truck in the yard. The woman wore a faded housedress" (8). Bizarrely, by using the very same techniques of fictional verisimilitude that Doctorow is deploying at that very moment (to allow us, the reader, to furnish the "storyworld" of *Andrew's Brain*), the character Andrew overreaches, achieving the opposite effect, undermining the "reality" of this fictional alter ego. The features on the map are *too* accurate for it not to be autobiographical. There are many such moments of playful inversion in the novel, as befits its underlying theme.

Now that he has broken cover, Andrew the cognitive scientist is openly distrustful of the therapy he has elected to undergo, his materialism reinforcing his skepticism. In one session, he recalls the "voices" that have been haunting him, an experience intimately linked to the act of writing, which, he says, is "like talking to yourself, which is what I have been doing all along with you, Doc" (49). More crucially, "I'm not sure we're talking about the same kind of experience. We have our manual too, you know. Your field is the mind, mine is the brain. Will the twain ever meet?" (12).

The fundamental schism is spelled out.

The Inadvertent Agent

Later, Doctorow will drop in speculation about artificial intelligence eclipsing human intelligence (post-humanism), conjecture about the eternal recurrence of the DNA molecule, the possibility of a collective intelligence at work in human choice, and much more.

Set against McEwan's linear thematic drive in *Saturday*, here, mediated by Andrew, the tone is tentative, unreliable, obsessive: where the top-down plot of *Saturday* comes across with all the force of a supertanker, *Andrew's Brain* is an unmoored skiff battered by opposing winds. To this end, the novel is constructed in almost Socratic form as an unresolved dialogue between two irreconcilable viewpoints. Yet, like McEwan, Doctorow finds the initial spark in his own memories.

> Many years ago, I worked for a man … [who] told me … that years before he had been feeding medicine to his infant child with an eye dropper and it was the wrong medication. And as a result, the child died. … And I also had the image in my mind—I don't know where it came from—of a girl with colored pencils drawing on a pad and then she sees an adult trying to see what she's done.
>
> And so she takes the pencil in her hand and scribbles over what she's been so carefully doing. And those two images somehow combined in some sort of evocative way and got me writing this book. (Online interview 2014b)[2]

So, chance and a strong sense of serendipity had a part in both the conception and realization of this work. Ideas about the brain may predominate here as in *Saturday*, but cognitive and neuroscientific issues are approached in quite contrasting ways.

Doctorow seemingly cherishes his own openness to creative unknowns, the "girl with colored pencils drawing on a pad" who, when observed by an adult, scribbles over her work. This personal recollection of his reflects warring creative forces: one aspect of the psyche spontaneous and impulsive, while the other coldly appraises with an "adult" or editorial eye.

He is undaunted by the fact that this image "came" to him, was in no way the result of conscious deliberation. As seen, the main plot-spring of the book also "came" to him, albeit in a different way. The defining event of Andrew's life, the accidental killing of the child, is something mined from Doctorow's own memory. There is a profound recognition here of his own lack of control over the material, "those two images [of the blighted man and the scribbling girl] somehow combined in *some sort of* evocative way" (our italics). With Henry Perowne, McEwan also makes use of personal memories, but in a much more conscious fashion, never quite ceding control to creative unknowns.

Elsewhere, Doctorow is even more unguarded: "Think of it as an installation. This book has a special coded way of working. It comes back on itself with its circulation of images or leitmotifs. ... This is a book that doesn't make a distinction between what might be real and what might not be real" (online interview 2014b).

And of course, that epistemological doubt about what we (and he) can know and what we cannot (what's "true" and what "came to him") comprises the book's core. Doctorow's blurring of the division between the "real" and the "unreal" suggests that the separation is often a spurious one, given the mind's insidious capacity for confabulation and false memory. As we're repeatedly told: "Pretending is the brain's work. It's what it does" (101). In that sense, he goes so far as to suggest, we're all suffering from "imposter syndrome." The book is an implicit recognition that this is true, too, of its author.

When pressed in interviews about the comparison to an "installation," he declines to elaborate, except by way of an apocalyptic vision of a post-fictional world:

> There are people building computers to emulate the brain and who believe in the theory that a computer can be achieved that has consciousness. ... If that ever happens, as Andrew assures us, it's the end of the mythic world we've lived in since the Bronze Age. The end of the Bible and all the stories we've told ourselves until now ... The impact of that would be equivalent to the planetary disasters of global warming, overpopulation or another giant asteroid blowing us to smithereens.
>
> (Online interview 2014b)

Clearly, behind this story of a hapless figure who turns everything he touches to dust lie massive themes. Doctorow, at the age of eighty-three, in what he suspects will be (and turned out to be) his last novel, is asking of us nothing less than a willingness to peer into the void at "the end of the mythic world." This apparently slight, quirky jeremiad about a crazed cognitive scientist quickly becomes an assault on some of our most treasured notions of literature, selfhood, and society.[3]

Toward the close, Andrew gets a chance, echoing the author's own words almost exactly, to spell out the precise meaning of the end of narrative to the two senior politicians scurrilously dubbed "Chaingang" and "Rumbum" (Dick Cheney and Donald Rumsfeld), one of the pleasures of this novel being Doctorow's eagerness to mix the puerile and the profound, the scurrilous and the scholarly.

> If we figure out how the brain gives us consciousness, we will have figured out how to replicate consciousness. You understand that, don't you Doc? I do. … Computers of course, I said, and animals genetically developed to have more than the primary consciousness of animals. To have feelings, states of mind, memory, longing. He means like in Disney, Rumbum said, and they laughed. I laughed as well. Yes, and with that all of that the end of the mythic world we've had since the Bronze Age. The end of our dominion. The end of the Bible and all the stories we've told ourselves until now.
>
> (186)

We are facing a revolution even greater than Crace's Bronze-age storyteller, since it involves the end of "writing" and "reading" as we know it. This is Doctorow speaking through his mouthpiece to warn us of the dangers ahead: a new postlapsarian era, where the fruit of the tree of knowledge now resides with other intelligences, artificial and genetically engineered; the effect so devastating that it marks no less than "the end of our dominion."

Elsewhere, Doctorow pointedly refers to the internet as the "overbrain," stressing the fact that this term is not some airy hypothetical. With his usual light touch, he also toys with the possibility that we are also subject to a government brain, the "brain" of collective memory, of IT "cloud thinking" (121,) or of our genetic legacy. *Andrew's Brain* is pervaded with a sense of humanity at a crossroads, or perhaps reaching the end of the line, and in that sense can be read as a parable.[4] This is an author contemplating the end of authors—and the end of readers in our current understanding of the term.

And what of us? Will we be able to empathetically relate to stories written by artificial intelligence, or by genetically enhanced animals? Such a dystopian state of affairs would also mean the end of all the "grand narratives" of religion. The death of the author would mean the death of God. Doctorow takes us to the edge of the abyss and makes us peer over.

Richard Kearney (see Introduction) is vocal in his warnings about the dangers of outsourcing ourselves—what he calls, "excarnation"—citing our recent tendency to live virtualized, disembodied lives. Doctorow, through Andrew, seems to pick up on this same anxiety, the irony being, of course, that Andrew himself is totally lost in his thoughts, adrift in his own body.

Guiltless yet Responsible

Andrew's "original sin," the defining action of his life, involves the accidental death of his daughter at his own hand: "In good faith I fed her the medicine I believed had been prescribed by our pediatrician. The druggist sent over the wrong medicine and I was not as alert as I should have been" (14).

Where, he demands, does ultimate responsibility lie? With himself? The druggist? A question that is all the more pertinent in light of some of Andrew's own concerns about free will: "I reacted badly to the publication of an experiment demonstrating that the brain can come to a decision seconds before we're conscious of it ... There will be more sophisticated experiments and it will be established that free will is an illusion" (32).[5]

It is as if Andrew finds himself guiltless yet responsible for his daughter's death, and all the other calamities that befall him. Without free will, can he really be held to account? Or, to resort to the terms of his own radical determinism, his brain may be guilty, but his conscious mind (and his "self") innocent.

Either way, such philosophical niceties mean little in the face of overwhelming grief. His defining catastrophe, though entirely inadvertent, Andrew takes upon himself. Moreover, he sees it as consistent with a pattern of personally initiated disasters which have tainted his entire life. In the course of the book, previous incidents emerge which seem to confirm his Job-like fate.

It's noteworthy that Andrew's crime—the one action from which all the others seem to derive—is a failure to read: "I could have read the label." Again, and quite deliberately, Doctorow places the act of reading and writing at the very fulcrum of the action in the novel. For this, surely, is Andrew's defining moment: the nightmare from which he cannot awake.

Yet there are many other "accidents" and errors he grimly relates, as if eager to confirm his pariah status. Some of these are horrific, like the death of his daughter, while others teeter on the edge of the comic macabre, such as the incident when he was a child and his dachshund is seized by a hawk. "It was all my fault. There you have it. Early Andrew" (19).

The calamities spill from him, one after the other.

He tells how, during a brief, enforced sojourn in the army, he fell asleep on nighttime maneuvers, and his platoon sergeant lost his stripes as a result. Later, at an academic party, he flung out his arm to make some point and inadvertently punched a "woman professor" sending her reeling to the floor. When her husband, enraged, chased after him, Andrew, startled, dropped a bottle of vodka on the man's foot, breaking it. "In the space of a minute I had taken out an entire family" (87).

Or the time when, as a boy, he was playing with his sled on the snowy sidewalk near his house, and decided to cross to the other side of the road, not seeing the approaching car. The car hits a pole, and the driver is killed, forcing Andrew's family to move from the area because of the bad feeling generated in the community: "My father said, Son, lots of kids were sleigh-riding and it could have been any one of them in the path of that car. It just happened to be you. He didn't believe this any more than I did" (56).

All these events, however superficially explicable, make Andrew feel like Typhoid Mary, or today's "superspreaders," inadvertently dispensing suffering into the lives of those rash enough to care for him. He stumbles through his own narrative like the epitome of entropy, destroying calm, upsetting order, generating "noise" in the system.

Martha, his first wife, is reduced to a shell of her former self after the death of their daughter, her career as a pianist prematurely ended. When Andrew himself finally falls in love again and marries the youthful Briony, it is not long before she herself is caught up in a catastrophe of global proportions—the twin towers—and disappears, presumed dead. "Nothing recognizably Briony was ever found" (131). Although Andrew bears no direct responsibility, by this stage he is so convinced of his malign influence that he even reads accusation in his new baby daughter's stare. "Willa would look at me with her mother's blue eyes. Quietly judgmental, I felt" (133). At this point, he makes the decision to give over Willa to his ex, Martha, in some barely conscious attempt to expiate his previous "crime," and to replace the daughter she had lost with him ("murdered"). Not far beneath the surface, this decision masked a more fundamental fear: "I'd killed one baby. [Do] you want me to kill another?" In regard to one of the most profound aspects of human life—parental care—Andrew's brain plainly no longer trusts Andrew's brain (18).

The novel as a whole could be read as an extended exercise in the capacity for "subjective validation," that is to say, our need to connect unrelated events, to fashion patterns from unpredictability. We are so acutely averse to the idea of randomness that we, or perhaps our left-brain "interpreters," are prepared to fashion convoluted justificatory narratives, overlooking flagrant improbabilities to do so.

For Andrew, the drive for subjective validation simply feeds his self-distrust, something that has further ramifications, given that he lectures in cognitive science. "One morning .... I asked this question: How can I think about my brain when it's my brain doing the thinking? So, is this brain pretending to be me thinking about it? I can't trust anyone these days, least of all myself." Once more, we seem to be peering into a bewildering hall of mirrors, undermining our most cherished certainties (32).

You Are Not in a Place

In another respect, the novel can be read as an exercise in playful philosophy of mind, with us as audience, looking on as our blighted avatar, Andrew, stumbles into one ontological mire after another. Here, Doctorow has pushed the postmodern device of the "unreliable narrator" to the limit by making him unreliable not just to us (the reader) but also to himself. Certainly, in the delusional torrent that is Andrew's stream of consciousness, there is an undertow of disengagement between "self" and "world." Talking to his class, he declares this openly.

> Consciousness .... [is when] you have nothing to fix on, nothing for your thought to adhere to, no image, no sound, no smell, no physical sensation of any kind. You are not in a place, you are the place. You are not here, you are everywhere. You are not in relation to anything else. There is no anything else. There is nothing you can think of except of yourself thinking.[6] (26)

What makes this vision more shocking is that he is describing basic sentience, not some aberrant state.

Very shortly after this passage, in an allusion to Plato's parable of the cave, Andrew, uncharacteristically emotional, draws a parallel between the narratively mediated nature of "normal" consciousness and what "happens in the heart" when you fall in love. Smitten by the beautiful acrobat Briony, he admits, that only now can he recognize life as it should be: "And what you thought of as life [was] only the shadows of the cave" (27).

What are we supposed to take from all this? That consciousness is a kind of hermetically-sealed chamber, or that love can set us free? It is entirely in keeping with Doctorow's overall approach that we are left on the back foot, confounded: for that is just where he wants us. Andrew's brain and indeed *Andrew's Brain* offer no easy answers, no nostrums, but merely the proliferation of questions so fundamental that we do not have the luxury of ignoring them.

But then this epistemological uncertainty ought not to be a surprise. Doctorow repeatedly reminds us of the brain's capacity for deception, including self-deception. Again, "Pretending is the brain's work ... The brain can even pretend not to be itself. Oh? What can it pretend to be, just by way of example? Well for the longest time, and until just recently, the soul" (102).

As a lecturer, Andrew enjoys teasing his students with unpalatable "truths" about human consciousness, as when he describes the paradox of evolved consciousness. "If consciousness exists without the world, it is nothing, and if it needs the world to exist, it is still nothing" (29).

Yet in the midst of these talks during which he invokes the spirits of "Emerson, William James, Damasio, and the rest" (29), Andrew retains (just) enough animal instinct, to find himself falling in love with Briony. His chosen means of seduction? Putting her in an EEG machine of course, flashing cards at her to see where the neural signal jumps. "I saw things more intimately Briony's than if I had seen her undressed. This wasn't mere voyeurism, it was cephalic-invasive" (31), a caricature of materialist reductionism and one in which Andrew gives a whole new dimension to the phrase "head-bound."

Taking our cue from Doctorow himself, and his gesture toward Damasio's work on embodied consciousness here, there are several points to make. Even as Andrew is intellectualizing his obsession with Briony, in this "invasive" experiment he is interacting with her physical being: her brain. This is seduction masquerading as science, a poor substitute for "see[ing] her undressed," the whole sorry episode a parody of the traditional Western insistence on creating a firewall between mind and body, the "high" and the "low," the spiritual and the carnal.

Damasio's writing has had many offshoots, one of which, carnal hermeneutics, championed by Richard Kearney, is focused on the seemingly irreconcilable dichotomy which torments and obsesses Andrew. If Damasio's mission is to convincingly portray the embodied mind, Kearney's is to philosophically reclaim from as far back as the ancient Greeks, embodied knowing, carnal intelligence.

Citing Keats' dictum that axioms in philosophy are not axioms until they are proved upon our pulses, Kearney emphasizes how many of our words for knowledge have somatosensory beginnings.[7] Indeed, he points out, even the "sapiens" that defines our species is derived from an etymological root that means "taste."

Savoir faire, being "savvy," have a similar basis—which is knowing through touching. "Tact" is developed from tactile. The word "sensible" and even the notion of "making sense" (as we've seen, one of the main functions of narrative) emerge by etymological extension from our corporeal senses.

Kearney's argument is that our ability to know anything derives not from reason, but from the totality of our physical being. Although building his assertions from Aristotle's claim in *De Anima* that touch is the primary sense, his more recent allies are those, like Damasio, who look at the mind/brain as an ensemble, an inextricably embodied entity, constructed from the ground up. To Damasio and Kearney, brain and body are indivisible, and even our sense of self emerges from constantly revised bodily feedback.

For those who subscribe to carnal hermeneutics, everything we discern about the world, we do through our bodies: "Life is hermeneutic through and through. It goes all the way up and all the way down. From head to foot and back again" (Kearney 2017: 24).[8] If true, this has huge connotations for all of us, but certainly for the beleaguered Andrew. It might well be that the profound schism between "the soul" and scientific materialism, mind and body he feels so acutely, is a spurious one.

Besieged by his demons, Andrew has no recourse to such a grounding realization. Resorting to his years of training, he tries to conceptualize away the defining crisis of his life—his daughter's death—to think his way out of the feelings of hurt and guilt he labors under. The unlikely vehicle for his release is a supercomputer. Suppose, he says to Briony, there was an immensely powerful supercomputer that could record not only the acts and thoughts and feelings of every person on earth, but also their memories.

What if such a computer could even store the evolutionary memory in each person's genome, asks Andrew. We begin to see where he is heading with this, "couldn't there be the possibility of recomposing a whole person from these bits and pieces and genomic memories of lives past?" He hurtles to a breathless conclusion, "then we could recover our lost babies, our lost lovers, our lost selves, bring them back from the dead, reunite in a kind of heaven on earth" (44).

Such a computer would be a resurrection machine. We, all our loved-ones could become Lazarus, liberated from death. Thus, in an act of desperation, or deep-seated self-delusion, Andrew summons up a technological salvation, a machine where time can be played backwards, his daughter rendered undead, and himself no longer a killer.

The scenario conjured up by Andrew, born of despair, is an extreme example of Kearney's "excarnation." At one level, it is a logical extension of the "brain in a vat" thought experiments proposed by philosophers in the last century. Damasio repeatedly ridicules the entire notion, stating that since mind derives from the entire organism; without the body, we are left with an empty collection of data, a digitalized pseudo-self. Once more, Andrew is seeking top-down, intelligence-driven solutions to problems arising from our innate and inescapable physicality.

Through the Looking-glass

In what is, first and foremost, a novel of ideas, Andrew's two loves—Martha and Briony—come across as conceptual proxies, or cyphers, rather than fully rounded figures. They are a means to investigate pressing issues, ranging from the eternal question of the connection between brain and mind, the embodied nature of consciousness, the problem of self-knowledge, right through to extreme, sometimes outlandish notions of collective intelligence, government as a "brain," post-humanism, and, in a solipsistic final turn, the mind as a prison: "It's a kind of jail, the brain's mind. We've got these mysterious three-pound brains and they jail us" (109).

Take Andrew's first sighting of Briony, performing gymnastics on the high bar with a group of other athletes. "The intensity, the concentration of each of them on what they were doing ... put him in mind of a culture of squiggling DNA molecules, so that if he waited long enough all these jumping and vaulting and circling squiggles would assemble themselves into the double helix of a genetic code" (21).

Clearly, she symbolizes our physical being, partly by reveling in the use of her own body, but, beyond this, reminds him of the genes that encode the body across generations. For Andrew, afflicted with chronic clumsiness, prone to knocking over lecterns, punching professors, and inflicting inadvertent flailing damage on those in his vicinity, she is a vision of easeful corporeality: his polar opposite, reminiscent of the earthy Strauss, playing squash against the cerebral Perowne. By this point, such polarities should not surprise us.

So lost is Andrew to his own feelings that it seems entirely appropriate his nickname is "Android." "I was Android all right. Tap me with your knuckle, hear the clunk" (170). At a later stage of events, in a paranoid vision, he even wonders, "Am I a computer? ... Am I the first computer invested with consciousness?" (195). Briony, on the other hand, is full of physical brio, even persuading him to go climbing with her. Of course, she strides ahead, vigorous and lithe, with him trailing admiringly behind, and they ascend "higher and higher, the path more like a series of cryptic Tibetan steps into a Buddhist acceptance of the way things really were when you didn't talk about them" (36).

Briony, all the while, is very much grounded—"sometimes touching the ground for support"—and yet, paradoxically, she leads Andrew upwards, out of his depression, "into a Buddhist acceptance," and even untrammeled joy. Yet even in the throes of infatuation, he cannot stop himself describing his feelings for her in terms of neurophysiology, "when I was with Briony there was nothing wrong with my limbic system. My hippocampus and my amygdala were up and running" (62).

Then there is the wonderful scene of Andrew meeting Briony's parents, retired show business people but also "diminutives" who had toured with various troupes of performing midgets. It is, of course, entirely in keeping with Briony's acrobatic nature that her parents should make a living based entirely on their own physical stature (66).

In a moment of strange inversion, he watches Briony embracing her mother as if the roles were reversed because of the size difference. Later, with the four of them at the dinner table, "Briony seemed huge ... if I stood up too suddenly I'd bang my head on the ceiling" (71). From his gargantuan vantage, Andrew endeavors to at least not trash the place, this once.

As is the way in this willfully quirky novel, key metaphors ripple and reverberate—the "phase music" sometimes adopting, sometimes abandoning a theme or motif. So, as regards elevation and debasement, the "high" and the "low," we are given the Wasatch mountains, scene of Briony's wooing (and also the Norwegian mountain cabin Andrew retreats to after her death), Briony's diminutive parents, the accident with the dog (a dachshund—you don't get much lower than that) ironically hoisted up into a tree and killed by the hawk. But also, in a humorous reversal of that predatory act, the sudden descent of a helicopter to embarrassingly intervene when Andrew and Briony are in the middle of lovemaking on a beach.

Being Andrew, he conceptualizes even this moment of uninhibited sensual release—reading into it a matter of "higher" and "lower" functions, resorting to a mental map entirely inappropriate to the situation. "It may begin as devotional, your lovemaking, but the brain goes dark, as a city blacks out, and some antediluvian pre-brain kicks in that all it knows to do is move the hips. It is surely some command from the Paleozoic Era, and may be the basis of all drumming" (82). Doctorow is riffing on simple reductionist notions of the brain and in this case (the triune brain, see Chapter 7) he has picked a worthy target.

The implication is that supposedly "higher" functions emerge from higher (i.e., more recent) strata of evolutionary development. Uncharacteristically, Andrew, burning with lust, is happy to descend into his "antediluvian pre-brain" and exist, however briefly, in an instinctual "Paleozoic," there on the dark beach, "and then, suddenly, there were blazing white lights above them, the unmistakable sound of a helicopter, swooping low, its rotors beating the air, its body resembling a black bug-like monster" (81).

Shocked by the violent intrusion, they scamper to their feet and flee.

Judgment comes from above in the form of an evolutionarily primitive form. Andrew, who normally exists at a remove from his own body, has for once dared to give way to its needs, and has been humiliated and condemned as a result. He is instantly cast out from his own "antediluvian" Eden.

The final, cataclysmic image of descent associated with Briony is, of course, one that altered the course of world history. The two of them have established themselves together in New York with their infant daughter, Willa, when Briony sees a documentary about the city's marathon. Watching the multitude of bodies jogging as one through the park, she is smitten with the idea. For his part, Andrew rationalizes her new fad as an example of the "collective brain" at work, "If I were ever to decide to do serious research in neuroscience—well, it would have to do with the communal brain. As with ants, as with bees" (119). He speculates further, "Perhaps we long for something like the situation these other creatures have … where the thinking is outsourced. Cloud thinking, a chemical übermensch" (121).

That morning, Briony sets off to train for her own marathon attempt, jogging down along the Hudson, after giving Andrew a brisk kiss on the cheek. It is the last time he will ever see her: "Nothing recognizably Briony was ever found" (131). Caught up in the hellish aftermath of the collapse of the twin towers, she is simply obliterated. There are no "remains." The body she was once so at home in leaves no trace whatsoever, but is instantly erased from time, unlike her young admirer Dirk who provides his own eerie epitaph in a recorded phone-message from the top of one of the burning towers before leaping to his death. We are poignantly reminded that Doctorow, in the act of writing his last novel, is in the process of consigning himself to his own version of posthumous reality: the mapmaker immortalizing himself in his chart.

Being and Becoming

Briony's death marks another major crux for Andrew. Thereafter he remains trapped in psychological limbo between the loss of his daughter, and this second abrupt bereavement. Throughout the novel, Doctorow weaves many clues that these two women represent not simply loss, but something more. In this respect, it is tempting to see Martha as "mind" and Briony (the name goads us) as "brain." Martha with her musicianship and access to "unconstrained emotions," and Briony, subject to EEG, the gymnast, the passionate lover, wholly embodied and replete in her physicality, "[who] was intensely physical, as grown children are" (37).

Andrew mischievously puts forward another antithesis. "Martha was being, Briony was becoming. What kind of a shrink are you who has to be told this?" (37). Indeed, he is caught for most of the book in a battle between these extremes, unable to integrate mind and brain, the mental and the physical, or indeed to find any kind of reconciliation between being and becoming. He exists in what sub-atomic physicists call an "agon-state," and the attempted resolution of his "agony" marks the real momentum of the narrative, such as it is, beneath all the obsessively reiterated leitmotifs, all the reverberant phase-music. In this sense, the book can be read as an eccentric *Pilgrim's Progress*, a neuroscientific parable of the twenty-first century, with Andrew as its wandering, penitential soul.

His torment is one of dis-integration, but also profound epistemological doubt. Hobbled by his own hyper-awareness of the brain's capacity for confabulation—"pretending is the brain's work"—he descends into paranoid delusion (101). Up on his Norwegian mountain, he is even tempted to dismiss the possibility of self-knowledge, "it is dangerous to stare into yourself. You pass through endless mirrors of self-estrangement. This too is the brain's cunning, that you are not to know yourself" (138).

The insidious effects of self-estrangement threaten to overwhelm what's left of his sanity on several occasions. He begins to imagine that he is in solitary confinement (another metaphor for the conscious mind) condemned to an endless Guantanamo-like incarceration, and that the shrink is actually one of his interrogators, "apart from talking to you I have no one … and I'm being held without trial and it's already been indefinitely" (193). In a novel this recursive, there can be no remission, no definitive release from the labyrinthine snares of "the brain's work."

Hand-stands and Somersaults

Yet there are moments of real epiphany. Significantly, one of these occurs after "Android" (now director of "Neurological research") lectures the politicians dubbed Chaingang and Rumbum on "neurological developments around the world"—specifically that the "great problem confronting neuroscience is how the brain becomes the mind. How the three-pound knitting ball makes you feel like a human being" (186). So, as in any good novel when his one climactic moment of rebellion arrives, it carries at once a sense of surprise and inevitability. First he tells Chaingang, Rumbum, and the president that they are prime examples of human insufficiency,

> Then I took a deep breath and did a handstand.
> A what?
> It just came over me, I was up on my hands almost before I realized it. Perhaps it was the image of Briony on the high bar—my first glimpse of her—that animated me, my brain having decided this was the thing to do, a mimetic act to bring her into resolution there in the White House ... I'd never done a bona fide hand-stand before. I was another man in the Oval Office.
>
> (187)

Unable to tell whether it is the example of Briony (or his brain) that leads him into this impulsive gesture, he simply knows that it feels right. He also knows that it will get him fired from his government position.

Yet this last "mimetic act" of the handstand, coming right at the end, also feels right to us, as readers. Then we think back to how Doctorow has primed us all along for exactly this moment, giving us not only the earlier sight of Briony on her high bar, but the woman in Prague who did a complete somersault in the air, head over heels "before landing nimbly on her bare feet" (72).

We cannot but think too of Briony's parents, the diminutives, performing acrobatic tricks in the circus. And then there is the image that Doctorow tells us was part of the genesis of the entire novel, of the girl in the farmhouse drawing with a crayon. When she sees an adult watching her, she "violently scribbles" all over her work, destroying it. Not until much later in the book is it revealed what she was working on. "It's all circus stuff. Acrobats, trapeze artists, tumblers, human pyramids" (158).

Andrew's handstand suddenly fits with a topsy-turvy pattern—of, if you like, putting the head below the body for once—which has already been established by Doctorow. All the "performers," Briony, her parents, and the Czech woman he once saw do a complete somersault represent a joy in physicality that seems to elude Andrew. Yet here, for one instant, in one subversive act, he is returned to himself, reacquiring the innocence of childhood. Andrew's child may be dead, but for a few seconds the child in Andrew lives on.

As mentioned, Doctorow explicitly points to the work of Antonio Damasio in the novel. *Descartes' Error* confronts precisely the false dichotomy between mind and body lived out by Andrew. As Damasio sees it, mind exists not at a remove, but is derived from a structural and functional ensemble of bodily constituents—from the entire organism. Mind is not top-down, but bottom-up, "neural circuits represent the organism continuously, as it is perturbed by stimuli from physical environments, and as it acts on those environments ... if the basic topic of those representations were not an organism anchored in the body, we might have some form of mind, but it would not be the mind we do have" (1994, 226). Or, as he pithily puts it: "no body, never mind."

Seen in this light, the repeated symbols of height and depth throughout the novel accrue new resonance. For, after all, what do we observe when we watch Andrew perform his handstand? The high brought low, and the low raised high. Even the "collective brain" of government is temporarily abased by his gesture. Then we look back not only at the influence of the various mountains on Andrew (the Wasatch peaks and the cabin in Norway), the ignominious end of the dachshund plucked up into the sky, the vision of Andrew and Briony looming over her "Lilliputian" parents, the sudden descent of the helicopter to interrupt their lovemaking, but also Briony's death, which although it was caused by a fall (of the twin towers) may have happened while she was rushing *up* "ninety five flights of stairs" to rescue her admirer, Dirk.

All these and other related images, like Doctorow's own version of the three-pound metaphor, "the three-pound knitting ball," serve to subvert our instinctive hierarchies, our sense of what is lofty and what is base, unsettling our tendency to place the "higher" mental faculties over "lower" bodily feelings, and return us to one of the few transcendent moments for Andrew, when the "great problem" at the core of the novel evaporates, and he exhibits incarnate, embodied consciousness at last.

I doubt if anyone had ever done a handstand in the Oval Office before. Really it was a triumph. I had for a moment risen out of my characteristic humility, my ordinary citizenness and in one upside down gesture achieved equity with these governors of my country ... And never again would I be another man according to the situation. I could feel my brain becoming me—we were resolved as one. (189)

Notes

1   http://bookpage.com/interviews/15792-e-l-doctorow#.VRkdGFboaRs.
2   http://bookpage.com/interviews/15792-e-l-doctorow#.VUHzJFboaRs.
3   There is, of course, a rich irony in "the end of story" being treated of as one of the themes in a story—another testimony to the flexibility and capaciousness of the novel.
4   Doctorow keeps returning to the notion that we are approaching various "tipping points" as a species—ecological and technological—and that the literary tradition within which he works is powerless in the face of these colossal forces.

5 This refers to a series of experiments by Benjamin Libet in the 1980s that suggested the brain's readiness potential precedes conscious decisions to perform *volitional* acts, implying that *unconscious* neuronal processes precede and even cause supposedly willed actions which are only retrospectively construed as consciously motivated. This is another kind of confabulation (or "pretending," in Andrew's words). Commentators have been debating the implications for free will ever since.

6 This is one of several points in the novel which seem to echo Lawrence Shainberg's 1988 novel about a neurosurgeon with altered subjectivity, *Memories of Amnesia.*

7 Keats, letter to John Reynolds, May 3, 1818. Kearney could as well have used another celebrated quotation from a letter of November 22, 1817 to Benjamin Bailey. "Oh, for a life of sensation rather than of thought."

8 See "Re-thinking the Senses"—lecture to University College Dublin, Newman Centre, January 23, 2017. https://www.youtube.com/watch?v=lWCqV8MAEpE Also, Kearney, Richard and Treanor, Brian, ed. 2015.

AFTERWORD

Yesterday upon the Stair

I Met a Man Who Wasn't There

For most of us, it is almost impossible not to complete the rest of the rhyme: testament to the power of memory. What's often lost is that the poem was written by a professor of pedagogy, Hughes Mearns (1955), whose life's work was encouraging the development of imagination in children. "He wasn't there again today/I wish, I wish he'd go away." Once completed, it is equally impossible not to apprehend, or at least engage with this ghostly, paradoxical entity—the man on the stair—despite everything logic tells us. Story demands it. Imagination will out.

As regards our exploratory narrative, we've come to the end of our journey. Certainly, there have been a few tangled jungles to traverse along the way, filled with strange flora and fauna, whose names we've had to learn as we went, together with some steep climbs. We also came across more than a few abandoned ruins, evidence of those who have come before.

We knew that we were entering contested space, one that the humanities and the sciences were both laying claim to in their different ways, but, confident that our purpose was exploratory rather than colonial, set off to see what we could find. One growing realization was that the gap between the two claimants is more apparent than real. The great anxiety in the humanist camp is that their entitlement to narrative, and even the very act of storytelling itself, will be supplanted, overturned. But no, literary criticism will simply reorient itself, something that is already taking place. The same goes for the specialized form of narrative that is literature. Writers of fiction are, in the main, not resisting this xenomorphic incursion, but accommodating, and even cultivating, it.

After all, the two camps may not speak the same language, but often have common interests, and can begin to resolve tensions in the time-honored way, by learning from one another and bartering ideas, new technologies, and maps written in their own particular styles of what the territory might contain. The maps made by one side may appear impressionistic, egocentric, and too subjectively topological to one, and needlessly obsessed with physical topography to the other. They can both be used.

Before written maps with elevation and scale, highly sophisticated "pre-lexical" Inuit and Aboriginal maps passed down orally, contained vital information, acquired over many generations of travel by foot. They were each ways of making sense of what William James called the great blooming, buzzing confusion of experience. For this enterprise we require both synthetic and analytic charts, brought together, "either by the agency of our own movements, carrying our senses from one part of space to another, or because new objects come successively to replace those by which we were at first impressed" (*Principles of Psychology*, 2018, 197). Not either, but also.

Some early travelers, like I. A. Richards, had already begun this barter, producing a beautifully detailed model of the reading process that relied upon, and was even ahead of, the neuroscience of his time. Starting with an acknowledgment that, as Jane Tompkins put it, the meaning of a poem or story has no existence, "outside its realization in the mind of the reader" (1980: ix) he began charting the relationship between cognition, emotion, and imagination.

We are stuck with our senses, and our remarkable but imperfect inner representations of outer space and events. Ultimately, subjectivity cannot be "analyzed away." This axiom must be acknowledged, and can be seen in much scientific research, in the recalcitrant presence of metaphor, imagination, even narrative itself. Of course the realm of subjective consciousness is where forms like the novel come into their own. Fiction grabs our attention and then channels it through a unique (albeit invented) personality. Point of view matters because I only have one. Yet, prompted by the writer, and our capacity for mentalizing, I can imaginatively extend myself by way of narrative, inhabiting the ventriloquist's dummy, seeing ghosts.

When we shift from noticing the textual cues and prompts such "channeling" involves, into how they achieve their effects—the details of neurophysiology and cognitive science—the move is in some ways analogous to that between worlding the story and storying the world. For, as we've seen, our outlook on reality and our small experience of it involves an innate "narrativity."

From both camps, there is a regrettable history of denial. Even two decades ago, study of the process of imagination was largely ignored by brain scientists, certainly by neuroscientists: that is, until 2007, when a series of reports linked the workings of memory and imagination—the ability to imaginatively place ourselves into possible futures. We don't just introject, we project. Imagination now demanded inclusion in the neuroscientific atlas.

Some literary critics, like Elaine Scarry, had already pre-empted this move. Using the available neuroscience of her day, she looked at reader response, how writers instruct, and our imagination, especially though visual means. Psychology too, came on board, examining how we form "situation models" as we read.

Given that reading and writing are recent cultural inventions, we have tried to include suggestion of some of the ways culture informs writing, and some of the observable ways it physically changes our brains. The skills of literacy actually involve a kind of retro-fitting of our basic neural equipment, re-adapting and changing a region of the brain originally used for edge and shape recognition, so that it can become a visual word form area (VWFA). We are not born with this, but create it in the process of learning to read—perhaps the ultimate example of neuroplasticity.

So these maps, deriving from one side or the other, are not definitive, but have to change as we do, and have to include a healthy respect for variation amongst individuals.

This fuels debate when it comes to emotion. The prized territory of the arts for many years, science has now much more to say. Some scientists link emotions to homeostatic survival mechanisms. But are emotions culturally inflected, or biologically given? More than a century before, Darwin identified what he saw as a set of emotional universals, such as joy and grief.

Others suggest instead that emotions are "psychologically constructed." This carries echoes of the old nature/nurture controversy. Perhaps we have now reached a point where we can agree there are some core emotions that all humans show, together with much cultural influence. Again, not either, but also.

Similarly, the notion that thought, and cognition more generally, is embodied is one that has excited many, while others remain unconvinced. It is helpful here to take a step back, something we have tried to do, and gain a fresh perspective, so as not to be overwhelmed by the nomenclature and methodology of one map or another. We need to ask, what does "embodied cognition" actually mean? We have used the term to describe the way our brain must interact with the world through the systems of our bodies. But do we think, feel, and create meaning in embodied ways? Almost certainly we do, but it is not the only way. We are still a symbol-making and using species.

There was a danger, in the clamor of such debate, of being distracted from our overall focus on narrative imagination. For this reason, we included two "field-studies" of fictional narratives which are informed by—and show a fascination with—the new information coming in across the old divide. *Saturday* and *Andrew's Brain* deal extensively with memory, emotion, and feelings, as you might argue all novels do, but they do so in the light of what they see as crucial re-drafting of the map of self, forcing us to accommodate a sometimes uncomfortable scientific take on what it means to be "me," even when, or especially when, that "me" is an entirely imagined creation. Their novels open up what can seem like esoteric debates into their implications: for the individual, for society, and for the grand project of storytelling itself. They also ask how far fiction can incorporate this new material, and whether it jeopardizes its very existence in doing so.

The answer is that stories will continue to enthrall us as long as there are people. As the study of the !Kung bushmen showed, entertainment is only one of its many functions. Salman Rushdie used a story to communicate with his son when he was in hiding, under death-threat. Scheherazade used a stream of them to survive. From the petty narratives of our daily lives, to the formalized narratives of science and literary theory, their production and reception are, as Jim Crace put it, "a necessary function of being human."

We come to the delicate issue of relative value. In the final analysis, which map is more trustworthy, more accurate, and which will have to give way in the sometimes heated negotiations between sides? We said near the start of this book that the interaction between narrative and neuroscience should be two-way. In answer to the abiding existential fear from the humanist side that story will be murdered by empricism, we would argue that, as of now, literature, and literary understanding, is offering much more to neuroscience than the converse. However, that may change.

It is fitting to end with the example of Darwin—that quintessential scientist—whose own struggle with narrative imponderables in his great theory caused him such anguish. He has given us an extraordinary late testimony of the effects of his life-long scientific labors upon his brain, which is also an inadvertent testament to neuroplasticity. Intriguingly, it also has direct bearing on his continued need for story and storytelling toward the end of his days. He relates how, up until the age of thirty or so, literature gave him "intense delight," but that latterly he "lost any taste" for it.

At the age of sixty-seven, looking back upon his life's work, he bewails the loss of these "higher tastes."

> My mind seems to have become a kind of machine for grinding general laws out of large collections of facts, but why this should have caused the atrophy of that part of the brain alone on which the higher tastes depend, I cannot conceive. ... The loss of these tastes is a loss of happiness.

Only one genre, it seems, holds out any hope of respite from this imaginative dearth—the novel. Ultimately, it is the novelist's art and—most importantly, the satisfaction of narrative closure—that Darwin most craves. Lacking to the end of his life the understanding of genes that would have conferred completeness on his own master plot, he gleans a vicarious consolation from the neatly resolved plots of others:

> Novels which are works of the imagination ... have been for years a wonderful relief and pleasure to me, and I often bless all novelists. A surprising number have been read aloud to me, and I like all if moderately good, and if they do not end unhappily—against which a law ought to be passed.

> (Darwin 1983: 83)

REFERENCES

Abbott, H. P. (2008), *The Cambridge Introduction to Narrative*, Cambridge UK: Cambridge University Press.

Abbott, H. P. (2013), *Real Mysteries*, Columbus OH: Ohio State University Press.

Abramson, L., H. Aviezer, I. Marom, and R. Petranker (2017), "Is fear in your head? A comparison of instructed and real-life expressions of emotion in the face and body," *Emotion*, 17:557–65.

Addis, D. R., A. T. Wong and D. L. Schacter (2007), "Remembering the past and imagining the future: Common and distinct neural substrates during event construction and elaboration," *Neuropsychologia*, 45:1363–77.

Adolphs, R. (2013), "The biology of fear," *Current Biol.*, 23:R79–R93.

Ahveninen, J., J. W. Belliveau, S. Huang, A. Y. Hung, I. P. Jääskeläinen, A. Nummenmaa, T. Raij, J. P. Rauschecker, S. Rossi, and H. Tiitinen (2013), "Evidence for distinct human auditory cortex regions for sound location versus identity processing," *Nature Commun.*, 4:2585.

Aidley, D. (1998), *The Physiology of Excitable Cells*, 4th edn, Cambridge UK: Cambridge University Press.

Aitchison, J. (1998), *The Articulate Mammal: An Introduction to Psycholinguistics*, 4th edn, Oxon UK: Routledge.

Albers, A. M., F. P. de Lange, H. C. Dijkerman, P. Kok, and I. Toni (2013), "Shared representations for working memory and mental imagery in early visual cortex," *Current Biol.*, 23:1427–31.

Alexander, W. and J. Brown (2019), "The role of the anterior cingulate cortex in prediction error and signaling surprise," *Topics in Cognitive Sci.*, 11:119–35.

Altmann, U., I. Bohm, A. M. Jacobs, O. Lubrich, and W. Menninghaus (2012), "The power of emotional valence—from cognitive to affective processes in reading," *Frontiers Human Neurosci.*, 6:192.

Amis, M. (1991), *Time's Arrow*, London: Jonathan Cape.

Anderson, Wendy (2015) [Interview] by L. Brooks, *The Guardian*, June 30. Available online: https://www.theguardian.com/books/2015/jun/30/metaphor-map-charts-the-images-that-structure-our-thinking

Anderson, A. K. and E. A. Phelps (2002), "Is the human amygdala critical for the subjective experience of emotion? Evidence of intact dispositional affect in patients with amygdala lesions," *J. Cognitive Neurosci.*, 14:709–20.

Andrews, M. (2014), *Narrative Imagination and Everyday Life*, Oxford UK: Oxford University Press.

Andrews-Hanna, J. R., R. L. Buckner, R. Poulin, J. S. Reidler, and J. Sepulcre (2010), "Functional-anatomic fractionation of the brain's default network," *Neuron*, 65:550–62.

Andrews-Hanna, J. R., R. Saxe, and T. Yarkoni (2014), "Contributions of episodic retrieval and mentalizing to autobiographical thought: Evidence from functional neuroimaging, resting state connectivity, and fMRI meta-analyses," *Neuroimage*, 91:324–35.

Apps, M. A., S. W. Chang, and M. F. Rushworth (2016), "The anterior cingulate gyrus and social cognition: Tracking the motivation of others," *Neuron*, 90:692–707.

Appenzeller, T. (2013), "Neanderthal culture: Old masters," *Nature*, 497:302–4.

Appleyard, B. (2011), *The Brain Is Wider than the Sky: Why Simple Solutions Don't Work in a Complex World*, London: Weidenfeld & Nicolson.

Arbib, M. A. (2003), *Handbook of Brain Theory and Neural Networks*, Cambridge MA: MIT Press.

Arbib, M. A. (2012), *How the Brain Got Language*, Oxford UK: Oxford University Press.

Arbib, M. A., V. Barres, and B. Gasser (2014), "Language is handy but is it embodied?" *Neuropsychologia*, 55:57–70.

Arevalo, A., J. Baldo, and N. Dronkers (2012), "What do brain lesions tell us about theories of embodied semantics and the human mirror neuron system?" *Cortex*, 48:242–54.

Armstrong, P. B. (2014), *How Literature Plays with the Brain: The Neuroscience of Reading and Art*, Baltimore MD: Johns Hopkins University Press.

Arnold, M. (1882), "Literature and science," The Rede Lecture, in *The Nineteenth Century*, Cambridge UK: Cambridge University Press.

Aubert, M., F. T. Atmoko, H. E. A. Brand, A. Brumm, D. L. Howard, J. Huntley, B. Istiawan, T. A. Ma'rifat, A. A. Oktaviana, E. W. Saptomo, P. Setiawan, P. H. Sulistyarto, P. S. C. Taçon, V. N. Wahyuono, and J. X. Zhao (2018), "Paleolithic cave art in Borneo," *Nature*, 564:254–7.

Aubert, M., R. Lebe, A. A. Oktaviana, M. Tang, B. Burhan, Hamrullah, A. Jusdi, Abdullah, B. Hakim, J-x. Zhao, I. Made Geria, P. Hadi Sulistyarto, R. Sardi and A. Brumm (2019), "Earliest hunting scene in prehistoric art," *Nature*, 576:442–5.

Auerbach, E. ([1953] 2003), *Mimesis: The Representation of Reality in Western Literature*, 50th Anniversary edn, Princeton NJ: Princeton University Press.

Austin, J. L. (1962), *How to Do Things with Words*, Cambridge MA: Harvard University Press.

Auyoung, E. (2018), *When Fiction Feels Real: Representation and the Reading Mind*, Oxford UK: Oxford University Press.

Aviezer, H., D. Gangi, W. Mattson, D. Messinger, A. Todorov, and S. Zangvil (2015), "Thrill of victory or agony of defeat? Perceivers fail to utilize information in facial movements," *Emotion*, 15:791–7.

Bajada, C., L. Cloutman, and M. L. Ralph (2015), "Transport for language south of the Sylvian fissure: The routes and history of the main tracts and stations in the ventral language network," *Cortex*, 69:141–51.

Baker, C. I., T. Benner, N. Kanwisher, K. K. Kwong, J. Liu, and L. L. Wald (2007), "Visual word processing and experiential origins of functional selectivity in human extrastriate cortex," *Proceedings Nat. Acad. Sci.*, 104:9087–92.

Bal, M. (2009), *Narratology: Introduction to the Theory of Narrative*, 3rd edn, Toronto ON: University of Toronto Press.

Bal, P. M. and M. Veltkamp (2013), "How does fiction reading influence empathy? An experimental investigation of the role of emotional transportation," *PLoS One*, 8(1):e55341.

Baldassano, C., U. Hasson, and K. A. Norton (2018), "Representation of real-world event schemas during narrative perception," *J. Neurosci.*, 38:9689–99.

Ballard, J. G. (1973), *Crash*, New York: Farrar Straus & Giroux.

Baron-Cohen, S. (1995), *Mindblindness: An Essay on Autism and Theory of Mind*, Cambridge MA: MIT Press.

Barrett, L. F. (2017), *How Emotions Are Made: The Secret Life of the Brain*, New York: Houghton Mifflin Harcourt.

Barrett, L. F. and A. B. Satpute (2013), "Large-scale brain networks in affective and social neuroscience: Towards an integrative functional architecture of the brain," *Current Opinion Neurobiol.*, 23:361–72.

Barry, P. (2002), *Beginning Theory: An Introduction to Literary and Cultural Theory*, 2nd edn, Manchester UK: Manchester University Press.

Bartlett, F. C. ([1932] 1995), *Remembering: A Study in Experimental and Social Psychology*, Cambridge UK: Cambridge University Press.

Bartolo, A., P. Baraldi, F. Benuzzi, P. Nicheli, and L. Nocetti (2006), "Humor comprehension and appreciation: An fMRI study," *J. Cognitive Neurosci.*, 18:1789–98.

Bateson, G. (2000), *Steps to an Ecology of Mind: Collected Essays in Anthropology, Psychiatry, Evolution, and Epistemology*, Chicago IL: University of Chicago Press.

Bauby, J. D. ([1997] 1998), *The Diving Bell and the Butterfly: A Memoir of Life in Death*, trans. J. Leggatt, New York: Vintage Books.

Beauchamp, M. (2015), "The social mysteries of the superior temporal sulcus," *Trends Cognitive Sci.*, 19:489–90.

Beckett, S. ([1931] 1999), *Proust and Three Dialogues with George Duthuit*, London: John Calder Publishers.

Beckett, S. (1979), *Company*, London: John Calder.

Beeman, M. J. (2005), "Bilateral brain processes for comprehending natural language," *Trends Cognitive Science*, 9:512–18.

Beeman, M. J., E. M. Bowden, and M. A. Gernsbacher (2000), "Right and left hemisphere cooperation for drawing predictive and coherence inferences during normal story comprehension," *Brain and Language* 71:310–36.

Belin, P., P. Ahad, B. Lafaille, B. Pike, and R. J. Zatorre (2000), "Voice-selective areas in human auditory cortex," *Nature*, 403:309–12.

Bernini, M. and A. Woods (2014), "Interdisciplinarity as cognitive integration: Auditory verbal hallucinations as a case study," *WIREs Cognitive Science*, September 1.

Berridge, K. C. and M. L. Kringelbach (2013), "Neuroscience of affect: Brain mechanisms of pleasure and displeasure," *Current Opinion Neurobiol.*, 23:294–303.

Binder, J. and R. Desai (2011), "The neurobiology of semantic memory," *Trends Cognitive Sci.*, 15:527–36.

Blakemore, C. (1977), *The Mechanics of the Mind*, Cambridge UK: Cambridge University Press.

Bogousslavsky, J. (2007), "Marcel Proust's lifelong tour of the Parisian neurological intelligentsia: From Brissaud and Dejerine to Sollier," *European Neurol.*, 57:129–36.

Bolhuis, J. J., R. C. Berwick, N. Chomsky, and I. Tattersall (2014), "How could language have evolved?," *PLoS Biol.*, 12:e1001934.

Borges, J. L. (1964), *Funes the Memorious*, in *Labyrinths*, trans. J. E. Irby, New York: New Directions Publishing Corporation.

Bower, G. and D. Morrow (1990), "Mental models in narrative comprehension," *Science*, 247:44–8.

Boyd, B. (2009), *On the Origin of Stories: Evolution, Cognition and Fiction*, Cambridge MA: Harvard University Press.

Boyd, B. (2018), "The evolution of stories: From mimesis to language, from fact to fiction," *WIREs Cogn. Sci.*, 9(1):e1444. doi: 10.1002/wcs.1444.

Brecht, M. (2007), "Barrel cortex and whisker-mediated behaviors," *Current Opinion Neurobiol.*, 17:408–16.

Broca, P. P. ([1865] 1986), "On the site of the faculty of articulated speech," trans. E. A. Berker, A. H. Berker, and A. Smith, *Archives Neurol.*, 43: 1065–72.

Brosch, R. (2018), "What we 'see' when we read: Visualization and vividness in reading fictional narratives," *Cortex*, 105:135–43.

Bruce, C., R. Desimone, and C. C. Gross (1981), "Visual properties of neurons in a polysensory area in superior temporal sulcus of the macaque," *J. Neurophysiol.*, 46:369–84.

Bruner, J. (1991), "The narrative construction of reality," *Critical Inquiry*, 18:1–21.

Bruner, J. (2004), "Life as narrative," *Social Research*, 71(3):691–708.

Buckner, R. L. (2000), "Neural origins of 'I remember'," *Nature Neurosci.*, 3:1068–9.

Buckner, R. L. (2012), "The serendipitous discovery of the brain's default network," *Neuroimage*, 62:1137–45.

Bullmore, E. (2016), "Why brains and airports have a lot in common," *BBC News* (health), February 17. Available online: http://www.bbc.com/news/health-35591696.

Butler, A. B. and W. Hodos (1996), *Comparative Vertebrate Neuroanatomy: Evolution and Adaptation*, New York: Wiley-Liss.

Buzsaki, G. and E. Moser (2013), "Memory, navigation and theta rhythm in the hippocampal-entorhinal system," *Nature Neurosci.*, 16:130–8.

Byars, S. G., D. Ewbank, D. R. Govindaraju, and S. C. Stearns (2010), "Natural selection in a contemporary human population," *Proceedings Nat. Acad. Sci.*, 107:1787–92.

Byatt, A. S. ([1994] 1998), *The Djinn in the Nightingale's Eye: Five Fairy Stories*, New York: Vintage, Random House.

Byatt, A. S. (2005), "Fiction informed by science," *Nature*, 434:294–7.

Byatt, A. S. (2006), "Observe the neurones: Between, above and below John Donne," *Times Literary Supplement*, September 22.

Byatt, A. S. (2008), "Introduction," in H. Harvey Wood and A. S. Byatt (eds), *Memory: An Anthology*, xii–xx, London: Chatto and Windus.

Byrne, R. W. and A. Whiten (1988), *Machiavellian Intelligence: Social expertise and the evolution of intellect in monkeys, apes and humans*, Oxford, UK: Clarendon Press.

Carter, C. S., D. Barch, M. Botvinick, T. Braver, J. D. Cohen, and D. Noll (1998), "Anterior cingulate cortex, error detection, and the online monitoring of performance," *Science*, 280:747–9.

Catania, K. C. and J. H. Kaas (1997), "Somatosensory fovea in the star-nosed mole: Behavioral use of the star in relation to innervation patterns and cortical representation," *J. Comparative Neurol.*, 387:215–33.

Cespedes-Guevara, J. and T. Eorola (2018), "Music communicates affects, not basic emotions—a constructionist account of attribution of emotional meanings to music," *Frontiers Psychol.*, 9:215.

Chang, E. F., M. S. Berger, and K. P. Raygor (2015), "Contemporary model of language organization: An overview for neurosurgeons," *J. Neurosurgery*, 122:250–61.

Chatman, S. (1980), *Story and Discourse: Narrative Structure in Fiction and Film*, Ithaca NY: Cornell University Press.

Chiou, R., G. Humphreys, J. Jung, and M. L. Ralph (2018), "Controlled semantic cognition relies upon dynamic and flexible interactions between the executive 'semantic control' and hub-and-spoke 'semantic representation' systems," *Cortex*, 103:100–16.

Cohen, L., S. Dehaene, S. Lehéricy, C. Lemer, O. Martinaud, M. Obadia, Y. Samson, and A. Slachevsky (2003), "Visual word recognition in the left and right hemispheres: anatomical and functional correlates of peripheral alexias," *Cerebral Cortex*, 13:1313–33.

Cohn, D. (1978), *Transparent Minds: Narrative Modes for Presenting Consciousness in Fiction*, Princeton NJ: Princeton University Press.

Collini, S. (1998), *Introduction to the Two Cultures* (Canto edn), Cambridge UK: Cambridge University Press.

Collini, S. (2013), "Leavis v Snow: The two-cultures bust-up 50 years on," *The Guardian*, August 16.

Conway, B. (2012), "Doing science, making art," *Trends Cognitive Sci.*, 16:310–12.

Corballis, M. (2007), "The Uniqueness of human recursive thinking," *American Scientist*, 95:240–8.

Corballis, M. (2015), *The Wandering Mind: What the Brain Does When You're Not Looking*, Chicago IL: University of Chicago Press.

Corballis, M. (2017), "The evolution of language: Sharing our mental lives," *J. Neurolinguistics*, 43:120–32.

Corballis, M. (2018), "Space, time, and language," *Cognitive Process.*, 19:89–92.

Corkin, S. ([2013] 2014), *Permanent Present Tense: The Man with No Memory and What He Taught the World*, London: Penguin Books.

Coslett, H. B. (2003), "Acquired Dyslexia," in K. Heilman and E. Valenstein (eds), *Clinical Neuropsychology*, 4th edn, 108–25, Oxford UK: Oxford University Press.

Courtiol, A., M. Jokela, V. Lummaa, J. E. Pettay and A. Rotkirch (2012), "Natural selection in a monogamous historical human population," *Proceedings Nat. Acad. Sci.*, 109:8044–9.

Courtiol, A., A. J. Fulford, V. Lummaa, A. M. Prentice, I. J. Rickard, and S.C. Stearns (2013), "The demographic transition influences variance in fitness and selection on height and BMI in rural Gambia," *Current Biol.*, 23:884–9.

Crace, J. (1988), *Gift of Stones*, New York: Charles Scribner's and Sons.

Crace, J. (2012), "[Interview] by J. Hodgson," *Review of Contemporary Fiction*, 32(3):1 September.

Craig, A. D. (2004), "Human feelings: Why are some more aware than others?" *Trends Cognitive Sci.*, 8:239–41.

Crish, S., K. Catania, C. Comer, and P. Marasco (2003), "Somatosensation in the superior colliculus of the star-nosed mole," *J. Comparative Neurol.*, 464:415–25.

Critchley, H. D., R. J. Dolan, A. Ohman, P. Rotshtein, and S. Wiens (2004), "Neural systems supporting interoceptive awareness," *Nature Neuroscience*, 7:189–95.

Crivellato, E. and D. Ribatti (2007), "Soul, mind, brain: Greek philosophy and the birth of neuroscience," *Brain Res. Bulletin*, 74:327–36.

Cummings, A., R. Ceponiene, F. Dick, A. Koyama, A. P. Saygin, and J. Townsend (2006), "Auditory semantic networks for words and natural sounds," *Brain Res.*, 1115:92–107.

Damasio, A. (1994), *Descartes Error: Emotion, Reason and the Human Brain*, New York: G.P. Putnam and Sons.

Damasio, A. ([1999] 2000), *The Feeling of What Happens: Body and Emotions in the Making of Consciousness*, London: Vintage.

Damasio, A. (2003), *Looking for Spinoza: Joy, Sorrow, and the Feeling Brain*, New York: Houghton Mifflin Harcourt.

Damasio, A. (2018), *The Strange Order of Things: Life, Feeling and the Making of Cultures*, New York: Pantheon.

Damasio, H., A. R. Damasio, R. Frank, A. Galaburda, and T. Grabowski (1994), "The return of Phineas Gage: Clues about the brain from the skull of a famous patient," *Science*, 264:1102–5.

Damasio, H., A. R. Damasio, T. J. Grabowski, R. D. Hichwa, and D. Tranel (1996), "A neural basis for lexical retrieval," *Nature*, 380:499–505.

Darwin, C. ([1859] 1945), *On the Origin of Species by Means of Natural Selection*, 6th edn, London: Watts and Co.

Darwin, C. ([1871] 1977), *Descent of Man and Selection in Relation to Sex*, New York: Modern Library.

Darwin, C. ([1872] 1965), *The Expression of the Emotions in Man and Animals*, Chicago IL: University of Chicago Press.

Darwin, C. ([1887] 1983), *Autobiographies: Charles Darwin and T.H. Huxley*, G. deBeer (ed), Oxford UK: Oxford University Press.

da Silva, S. G. and J. J. Tehrani (2016), "Comparative phylogenetic analyses uncover the ancient roots of Indo-European folktales," *Royal Society open science*, 3:150645.

Davey, J., A. Costigan, K. Kreiger-Redwood, N. Murphy, and S. A. Reuschemeyer (2015), "Shared neural processes support semantic control and action understanding," *Brain & Language*, 142:24–35.

Davies, R. (1990), *A Voice from the Attic: Essays on the Art of Reading*, Penguin revised edn, New York: Penguin Books.

Davis, P. (2007), *Shakespeare Thinking*, London: Continuum Books.

Dayan, P. and L. F. Abbott (2005), *Theoretical Neuroscience: Computational and Mathematical Modeling of Neural Systems*, Cambridge MA: MIT Press.

Decety, J. (1994), "Mapping motor representations with positron emission tomography," *Nature*, 371:600–2.

Deen, B., N. Kanwisher, K. Koldewyn, and R. Saxe (2015), "Functional organization of social perception and cognition in the superior temporal sulcus," *Cerebral Cortex*, 25:4596–609.

Deen, B., D. D. Dilks, N. Kanwisher, B. Keil, H. Richardson, R. Saxe, A. Takahashi, and L. L. Wald (2017), "Organization of high-level visual cortex in human infants," *Nature Commun.*, 8:13995.

Dehaene, S. (2009), *Reading in the Brain: The New Science of How We Read*, London: Penguin Books.

Dehaene, S. (2013), "Inside the letterbox: How literacy transforms the human brain," *Cerebrum*, 7:1–16.

Dehaene, S. and L. Cohen (2007), "Cultural recycling of cortical maps," *Neuron*, 56:384–98.

Dehaene, S., L. Cohen, R. Kolinsky, and J. Morias (2015), "Illiterate to literate: Behavioural and cerebral changes induced by reading acquisition," *Nature Revs. Neurosci.*, 16:234–44.

Deniz, F., J. Gallant, A. Huth, and A. Nunez-Elizalde (2019), "The representation of semantic information across human cerebral cortex during listening versus reading is invariant to stimulus modality," *J. Neurosci.*, 39:7722–36.

Dennett, D. C. ([1995] 1996), *Darwin's Dangerous Idea: Evolution and the Meanings of Life*, Touchstone edn, New York: Simon and Schuster.

Deresiewicz, W. (2012), *A Jane Austen Education: How Six Novels Taught Me about Love, Friendship, and the Things That Really Matter*, New York: Penguin Books.

De Voogd, L., G. Fernandez, and E. Hermans (2016), "Disentangling the roles of arousal and amygdala activation in emotional declarative memory," *Social Cognitive & Affective Neurosci.*, 11:1471–80.

Dick, A. S., B. Bernal, and P. Tremblay (2014), "The language connectome: New pathways, new concepts," *The Neuroscientist*, 20:453–67.

Dijkstra, N., S. Bosch, and M. van Gerven (2017), "Vividness of visual imagery depends on the neural overlap with perception in visual areas," *J. Neuroscience*, 37:13767–73.

Dillard, A. (1989), *The Writing Life*, New York: Harper and Row.

Doctorow, E. L. (2014a), *Andrews Brain*, London: Little, Brown.

Doctorow, E. L. (2014b), [Online interview] by A. Mudge, *Bookpage*, January. Available online: http://bookpage.com/interviews/15792-e-l-doctorow#.VRkdGFboaRs

Doctorow, E. L. (2014c), [Radio interview] "Doctorow ruminates on how a 'brain' becomes a mind," N.P.R., January 11.

Donald, M. (1991), *Origins of the Modern Mind: Three Stages in the Evolution of Culture and Cognition*, Cambridge MA: Harvard University Press.

Dor, D. (2015), *The Instruction of Imagination: Language as a Social Communication Technology*, New York: Oxford University Press.

Dor, D. (2017), "From experience to imagination: Language and its evolution as a social communication technology," *J. Neurolinguistics*, 43:107–19.

Dronkers, N. F., J. Jaeger, B. Redfern, R. Van Alein, and D. Wilkins (2004), "Lesion analysis of the brain areas involved in language comprehension," *Cognition*, 92:145–77.

Dronkers, N. F., O. Plaisant, M.T. Iba-Zizen, and E. A. Cabanis (2007), "Paul Broca's historic cases: High resolution MR imaging of the brains of Leborgne and Lelong," *Brain*, 130:1432–41.

Dunbar, R. I. M. (1998), "The social brain hypothesis," *Evol. Anthropology*, 6:178–90.

Dunbar, R. I. M. (2003), "Evolution of the social brain," *Science*, 302:1160–61.

Dunbar, R. I. M. and S. Shultz (2007), "Evolution in the social brain," *Science*, 317:1344–7.

Eagleman, D. (2011), *Incognito: The Secret Lives of the Brain*, New York: Vintage Books.

Eagleton, T. (2002), "A good reason to murder your landlady," *London Review of Books*, 24(8).

Eakin, P. J. (1999), *How Our Lives Become Stories: Making Selves*, Ithaca NY: Cornell University Press.

Easterlin, N. (2012), *A Biocultural Approach to Literary Theory and Interpretation*, Baltimore MD: Johns Hopkins University Press.

Edelman, G. M. and G. Tononi (2001), *A Universe Of Consciousness: How Matter Becomes Imagination*, New York: Basic Books.

Ekman, P., A. Chan, I. Diacoyanni-Tarlatzis, W. V. Friesen, K. Heider, R. Krause, W.A. LeCompte, M. O'Sullivan, T. Pitcairn, and P. E. Ricci-Bitti (1987), "Universals and cultural differences in the judgements of facial expressions and emotion," *J. Personality Social Psychol.*, 53:712–17.

Elbert, T., C. Pantev, B. Rockstroh, E. Taub, and C. Weinbruch (1995), "Increased cortical representation of the fingers of the left hand in string players," *Science*, 270:305–7.

Eldridge, L. L., S. Y. Bookheimer, S. A. Engel, C. S. Furmanski, and B. J. Knowlton (2000), "Remembering episodes: A selective role for the hippocampus during retrieval," *Nature Neurosci.*, 3:1149–52.

Emmott, C. ([1997] 1999), *Narrative Comprehension: A discourse perspective*, Oxford UK: Oxford University Press.

Engström, M., A. D. Craig, T. Karlsson, and A. M. Landtblom (2015), "Evidence of conjoint activation of the anterior insular and cingulate cortices during effortful tasks," *Frontiers Human Neurosci.*, 8:1071.

Esopenko, C., R. Borowsky, J. Cummine, L. Gould, N. Kuhlmann, and G. E. Sarty (2012), "A neuroanatomical examination of embodied cognition: Semantic generation of action-related stimuli," *Frontiers Human Neurosci.*, 6:84.

Ewert, J.- P. (1997), "Neural correlates of key stimulus and releasing mechanism: A case study of two concepts," *Trends Neurosci.*, 20:332–9.

Farah, M. J. (2003), "Disorders of visual-spatial perception and cognition," in K. M. Heilman and E. Valenstein (eds), *Clinical Neuropsychology*, 4th edn, chapter 8, Oxford UK: Oxford University Press.

Fauconnier, G. and M. Turner (2002), *Conceptual Blending and the Mind's Hidden Complexities*, New York: Basic Books.

Faust, M., D. Balota, J. Duchek, and M. A. Gernsbacher (1997), "Inhibitory control during sentence comprehension in individuals with Dementia of the Alzheimer type," *Brain & Language*, 57:225–53.

Favorini, A. (2011), "Memory in theater: The scene is memory," in M. Matthews, J. L. McClelland and S. Nalbantian (eds), *The Memory Process: Neuroscientific and Humanistic Perspectives*, 315–34, Cambridge MA: MIT Press.

Feinstein, J. S. (2013), "Lesion studies of human emotion and feeling," *Current Opinion Neurobiol.*, 23:304–9.

Fish, S. (1967), *Surprised by Sin: The Reader in Paradise Lost*, London: Macmillan.

Fish, S. (1970), "Literature in the reader: Affective stylistics," *New Literary History*, 2:123–62.

Fish, S. (1982), *Is There a Text in This Class? The Authority of Interpretive Communities*, Cambridge MA: Harvard University Press.

Fish, S. (2007), "Why do writers write?" *New York Times*, February 11. Available online: http://opinionator.blogs.nytimes.com/2007/02/11/why-do-writers-write/

Forseth, K., C. Conner, G. Hickok, C. Kaddiapasaoglu, R. Knight, and N. Tandon (2018), "A lexical semantic hub for heteromodal naming in middle fusiform gyrus," *Brain*, 141:2112–26.

Foxwell, J., Alderson-Day, B., Fernyhough, C., and Woods, A. (2020), "I've learned I need to treat my characters like people': Varieties of agency and interaction in writers' experiences of their characters' voices," *Consciousness and Cognition*, 79: 102901.

Friederici, A. D. (2006), "What's in control of language?" *Nature Neuroscience* 9:991–2.

Friederici, A. D., B. Opitz, and D. Yves von Cramon (2000), "Segregating semantic and syntactic processing in the human brain: An fMRI investigation of different word types," *Cerebral Cortex*, 10:698–705.

Gannon, P. J., A. R. Braun, D. C. Broadfield, and R. L. Holloway (1998), "Asymmetry of Chimpanzee planum temporale: Humanlike pattern of Wernicke's brain language area homolog," *Science*, 279:220–2.

Garfinkel, S. N. and H. D. Critchley (2016), "Threat and the body: How the heart supports fear processing," *Trends Cognitive Sci.*, January 20:34–46.

Gazzaniga, M. S. (2000), "Cerebral specialization and interhemispheric communication. Does the corpus callosum enable the human condition?" *Brain*, 123:1293–326.

Gazzaniga, M. S. (2005), "Forty-five years of split-brain research and still going strong," *Nature Revs. Neurosci.*, 6:653–9.

Gazzaniga, M. S. (2013), "Shifting gears: Seeking new approaches for mind/brain mechanisms," *Annual Rev. Psychol.*, 64:1–20.

Gelbard-Sagiv, H., I. Fried, M. Harel, R. Malach, and R. Mukamel (2008), "Internally-generated reactivation of single neurons in human hippocampus during free recall," *Science*, 322:96–101.

Gelernter, D. (2014), "The closing of the scientific mind," *Commentary*, 137(1):17–25.

Gernsbacher, M. A. (1997), "Two decades of structure building," *Discourse Processes*, 23:265–304.

Gernsbacher, M. A., H. H. Goldsmith, and R. Robertson (1992), "Do readers mentally represent characters' emotional states?" *Cognition & Emotion*, 6:89–111.

Gerrig, R. (1993), *Experiencing Narrative Worlds: On the Psychological Activities of Reading*, New Haven CT: Yale University Press.

Gerrig, R. and M. Jacovina (2009), "Reader participation in the experience of narrative," *Psychol. of Learning & Motivation*, 51(7):223–54.

Geshwind, N. and W. Levitsky (1968), "Left-right asymmetries in temporal speech region," *Science*, 161:186–7.

Goldie, P. (2012), *The Mess Inside: Narrative, Emotion, & the Mind*, Oxford UK: Oxford University Press.

Golding, W. G. ([1955] 1988), *The Inheritors*, London: Faber & Faber.

Golding, W. G. (1982), *A Moving Target*, London: Faber & Faber.

Goldstein, R. ([1983] 1993), *The Mind-Body Problem*, New York: Penguin Books.

Gonzalez, J., C. Avila, A. Barros-Loscertales, V. Belloch, V. Meseguer, F. Pulvermüller, and A. Sanjuán (2006), "Reading *cinnamon* activates olfactory brain regions," *Neuroimage*, 32:906–12.

Goodblatt, C. and J. Glicksohn (2010), "Conversations with I. A. Richards: The renaissance in cognitive literary studies," *Poetics Today*, 33:387–432.

Gould, S. J. (2003), *The Hedgehog, the Fox, and the Magister's Pox*, New York: Harmony Books, Three Rivers Press.

Graves, R. (1973), *Difficult Questions, Easy Answers*, Garden City NY: Doubleday & Co.

Gregg, M. and G. Seigworth (2010), *The Affect Theory Reader*, Durham NC: Duke University Press.

Grobstein, P. (2005), "Revisiting science in culture: Science as story telling and story revising," *J. Res. Practice*, 1(1):article M1.

Grobstein, P. (2007), "Interdisciplinarity, transdisciplinarity and beyond: The brain, story sharing, and social organization," *J. Res. Practice*, 3(2):article M21.

Grobstein, P., C. Comer, and S. Kostyk (1983), "Between sensory maps and directed motor output," in J. P. Ewert, R. R. Capranica, and D.J. Ingle (eds), *Advances in Vertebrate Neuroethology*, 331–47, New York: Plenum Press.

Gross, C. G. (2002), "Genealogy of the grandmother cell," *The Neuroscientist*, 8:512–18.

Gross, C. G. (2008), "Single neuron studies of inferior temporal cortex," *Neuropsychologia*, 46(3):841–52.

Haddon, M. (2004), *The Curious Incident of the Dog in the Night-Time*, New York: Vintage Press.

Hagmann, P., L. Cammoun, X. Gigandet, R. Meuli, C. J. Honey, V. J. Wedeen, and O. Sporns. (2008), "Mapping the structural core of human cerebral cortex," *PLoS Biology* 6(7):e159.

Hagoort, P. (2008), "The fractionation of spoken language understanding by measuring electrical and magnetic brain signals," *Philosophical Trans. Royal Society Lond.*, B363:1055–69.

Hales, D. (2009), *La Bella Lingua*, New York: Broadway Books.

Halliday, M. A. K. ([1971] 2002), *Linguistic Studies of Text and Discourse*, J. Webster (ed), London: Continuum.

Hari, R., L. Henrickson, S. Malinen, and L. Parkkonen (2015), "Centrality of social interaction in human brain function," *Neuron*, 88:181–93.

Harris, P. L. (2000), *The Work of the Imagination*, Oxford UK: Wiley-Blackwell.

Hartzell, J. F., B. Davis, D. Melcher, G. Miceli, J. Jovicich, T. Nath, N. C. Singh, and U. Hasson (2016), "Brains of verbal memory specialists show anatomical differences in language, memory and visual systems," *Neuroimage*, 131:181–92.

Harvey Wood, H. and A. S. Byatt (2008), *Memory, An Anthology*, London: Chatto and Windus.

Hassabis, D., D. Kumaran, E. A. Maguire, and S. D. Vann (2007), "Patients with hippocampal amnesia cannot imagine new experiences," *Proceedings Nat. Acad. Sci.*, 104:1726–31.

Hauk, O., I. Johnsrude, and F. Pulvemuller (2004), "Somatotopic representation of action words in human motor and premotor cortex," *Neuron*, 41:301–7.

Hauser, M. D., N. Chomsky, and W. T. Fitch (2002), "The faculty of language: What is it, who has it, and how did it evolve?" *Science*, 298:1569–79.

Hebb, D. O. (1949), *Organization of Behavior*, New York: John Wiley & Sons.

Heberlein, A., R. Adolphs, H. Damasio, and D. Tranel (2004), "Cortical regions for judgments of emotions and personality traits from point-light walkers," *J. Cognitive Neurosci.*, 16:1143–58.

Henrich, J., S. J. Heine, and A. Norenzayan (2010), "The weirdest people in the world?" *Behavioral Brain Sci.*, 33:61–135.

Henshilwood, C. S., F. d'Errico, G. A. Duller, Z. Jacobs, D. Mercier, J. C. Sealy, C. Tribolo, H. Valladas, I. Watts, A. G. Wintle, and R. Yates (2002), "Emergence of modern human behavior: Middle stone age engravings from South Africa," *Science*, 295:1278–80.

Henshilwood, C. S., Y. Coquinot, F. d'Errico, R. García-Moreno, Z. Jacobs, S. E. Lauritzen, M. Menu, and K. L. van Niekerk (2011), "A 100,000-year-old ochre-processing workshop at Blombos Cave, South Africa," *Science*, 334:219–22.

Henshilwood, C. S., L. Dayet, F. d'Errico, L. Pollarolo, A. Queffelec, and K. L. van Niekerk (2018), "An abstract drawing from the 73,000-year-old levels at Blombos Cave, South Africa," *Nature*, 562:115–18.

Herbert, C., S. Andres, T. Erhofer, W. Grodd, M. Junghofer, J Kissler, and D. Wildgruber (2009), "Amygdala activation during reading of emotional adjectives—an advantage for pleasant content," *Social Cognitive & Affective Neurosci.*, 4:35–49.

Herman, D., ed. (2003), *Narrative Theory and the Cognitive Sciences*, Stanford CA: CSLI publications.

Herman, D. (2013), *Storytelling and the Science of the Mind*, Cambridge MA: MIT Press.

Heyward, F. D. and J. D. Sweatt (2015), "DNA methylation in memory formation: Emerging insights," *The Neuroscientist*, 21:475–89.

Hickok, G. (2014), *The Myth of Mirror Neurons: The Real Neuroscience of Communication and Cognition*, New York: W. W. Norton and Co.

Hickok, G. and D. Poeppel (2007), "The cortical organization of speech processing," *Nature Revs. Neuroscience*, 8:393–402.

Hillis, A. E. and B. C. Rapp (2004), "Cognitive and neural substrates of written language: Comprehension and production," in M. S. Gazzaniga (ed), *The Cognitive Neurosciences III*, 775–87, Cambridge MA: MIT Press.

Hirst, W., R. L. Buckner, A. E. Budson, A. F. Cuc, J. D. Gabrieli, M. K. Johnson, C. Lustig, K. B. Lyle, M. Mather, R. Meksin, K. J. Mitchell, K. N. Ochsner, A. Olsson, E. A. Phelps, D. Schacter, J. S. Simons, and C. J. Vaidya (2015), "A ten-year follow-up of a study of memory for the attack of September 11, 2001: Flashbulb memories and memories for flashbulb events," *J. Exp. Psychol. Gen.*, 144(3):604–23.

Hirstein, W. (2006), *Brain Fiction: Self-Deception and the Riddle of Confabulation*, Cambridge MA: MIT Press.

Hoffmann, R. (2012), *Roald Hoffmann on the Philosophy, Art, and Science of Chemistry*, Oxford UK: Oxford University Press.

Hogan, P. C. (2003a), *The Mind and Its Stories*, Cambridge UK: Cambridge University Press.

Hogan, P. C. (2003b), *Cognitive Science Literature and the Arts: A Guide for Humanists*, New York: Routledge.

Hogan, P. C. (2010), "Literary universals," in L. Zunshine (ed), *Introduction to Cognitive Cultural Studies*, 37–60, Baltimore MD: Johns Hopkins University Press.

Hogan, P. C. (2011a), *What Literature Teaches Us about Emotion*, Cambridge UK: Cambridge University Press.

Hogan, P. C. (2011b), *Affective Narratology*, Lincoln NE: University of Nebraska Press.

Hogan, P. C. (2013), *How Author's Minds Make Stories*, Cambridge UK: Cambridge University Press.

Holland, N. N. (2009), *Literature and the Brain*, Gainesville FL: The PsyArt Foundation.

Hooper, J. and D. Teresi (1986), *The 3-Pound Universe*, New York: Tarcher/Perigee Books.

Hubel, D. and T. N. Wiesel (1968), "Receptive fields and functional architecture of monkey striate cortex," *J. Physiol.*, 195:215–43.

Hubel, D. and T. N. Wiesel (1977), "Ferrier lecture. Functional architecture of macaque monkey visual cortex," *Proceedings Royal Society Lond.*, B198:1–59.

Huey, E. B. ([1908] 1968), *The Psychology and Pedagogy of Reading*, Cambridge MA: MIT Press.

Hulme, C. and M. J. Snowling (2014), "The interface between spoken and written language: Developmental disorders," *Philosophical Trans. Royal Society* Lond., B369:20120395.

Huth, A., J. Gallant, T. Griffiths, W. de Heer, and F. Theunissen (2016), "Natural speech reveals the semantic maps that tile human cerebral cortex," *Nature*, 532:453–8.

Huxley, A. (1963), *Literature and Science*, Woodbridge CT: Ox Bow Press.

d'Huy, J. (2016), "The evolution of Myths," *Scientific American*, 315(6):62–9.

Ingle, D. (1973), "Two visual systems in the frog," *Science*, 181:1053–5.

Iredale, J. M., A. Dimoska-Marco, S. McDonald, J. A. Rushby, and J. Swift (2013), "Emotion in voice matters: Neural correlates of emotional prosody perception," *Int. J. Psychophysiol.*, 89:483–90.

Iser, W. (1980), *The Act of Reading: A Theory of Aesthetic Response*, Baltimore MD: Johns Hopkins University Press.

Iser, W. (1993), *The Fictive and the Imaginary: Charting Literary Anthropology*, Baltimore MD: Johns Hopkins University Press.

Ison, M. J., I. Fried, and R. Quian Quiroga (2015), "Rapid encoding of new memories by individual neurons in the human brain," *Neuron*, 87:220–30.

Izard, C. E. (1994), "Innate and universal facial expressions: Evidence from developmental and cross-cultural research," *Psychol. Bulletin*, 115:288–99.

Jack, R. E., R. Caldara, O. G. B. Garrod, P. G. Schyns, and H. Yu (2012), "Facial expressions of emotion are not culturally universal," *Proceedings Nat. Acad. Sci.*, 109:7241–4.

Jack, R. E., R. Caldara, O. G. B. Garrod, P. G. Schyns, and H. Yu (2013), "Reply to Sauter and Eisner: Differences outweigh commonalities in the communication of emotions across cultures," *Proceedings Nat. Acad. Sci.*, 110:E181–E182.

Jackendoff, R. (2002), *Foundations of Language: Brain, Meaning, Grammar, Evolution*, Oxford UK: Oxford University Press.

Jackendoff, R. (2012), *A User's Guide to Thought and Meaning*, Oxford UK: Oxford University Press.

Jackson, J. C., J. Watts, T. R Henry, J-M. List, R. Forkel, P. J. Mucha, S. J. Greenhill, R. D. Gray and K. A. Lindquist (2019), "Emotion semantics show both cultural variation and universal structure," *Science*, 366:1517–22.

James, W. ([1890] 1950), *Principles of Psychology, Volume Two*, Mineola NY: Dover Publications.

Jang, G., S. Yoon, J. Kim, J. H. Ko, S. Lee, H. Park, and H. J. Park (2013), "Everyday conversation requires cognitive inference: Neural bases of comprehending implicated meanings in conversations," *Neuroimage*, 81:61–72.

Jasmin, K., C. Lima, and S. Scott (2019), "Understanding rostral—caudal auditory cortex contributions to auditory perception," *Nature Revs. Neurosci.*, 20:425–34.

Jeffries, E. (2013), "The neural basis of semantic cognition: Converging evidence from neuropsychology, neuroimaging and TMS," *Cortex*, 49:611–25.

Jenkins, W. M., T. Allard, E. Guic-Robles, M. M. Merzenich, and M. T. Ochs (1990), "Functional reorganization of primary somatosensory cortex in adult owl monkeys after behaviorally controlled tactile stimulation," *J. Neurophysiol.*, 63:82–104.

Johnson, D. R. (2012), "Transportation into a story increases empathy, prosocial behavior, and perceptual bias toward fearful expressions," *Personality Individual Diffs*, 52:150–5.

Johnson-Laird, P. N. (1983), *Mental Models: Towards a Cognitive Science of Language, Inference, and Consciousness*, Cambridge MA: Harvard University Press.

Jones, G. (1993), "A protoconnectionist theory of memory," *Memory & Cognition*, 21:375–78.

Josselyn, S. A., P. W. Frankland, and S. Köhler (2015), "Finding the engram," *Nature Revs. Neurosci.*, 16(9):521–34.

Joyce, J. ([1922] 1960), *Ulysses*, London: Penguin.

Kandel, E. R. (2006), *In Search of Memory: The Emergence of a New Science of Mind*, New York: W. W. Norton and Co.

Kandel, E. R. (2012), *The Age of Insight: The Quest to Understand the Unconscious in Art, Mind and Brain, from Vienna 1900 to the Present*, New York: Random House.

Kandel, E. R., A. J. Hudspeth, T. M. Jessel, J. H. Schwartz, and S. A. Siegelbaum, eds (2013), *Principles of Neural Science*, 5th edn, New York: McGraw-Hill Companies, inc.

Kanwisher, N. (2006), "Neuroscience. What's in a face?" *Science*, 311:617–18.

Karadamun, A., A. Chatterjee, and T. Göksun (2017), "Narratives of focal brain injured individuals: A macro-level analysis," *Neuropsychologia*, 99:314–25.

Kearney, R. (2002) *On Stories*, London: Routledge.

Kearney, R. (2017), "Narrative and recognition in the flesh," [Online interview] by M. Concalo, *Philosophy and Social Criticism*, January 31. Available online: https://journals.sagepub.com/doi/10.1177/0191453716688367

Kearney, R. and B. Treanor, eds (2015), *Carnal Hermeneutics*, New York: Fordham University Press.

Keidel, J. L., P. M. Davis, V. Gonzalez-Diaz, C. D. Martin, and G. Thierry (2013), "How Shakespeare tempests the brain: Neuroimaging insights," *Cortex*, 49(4):913–19.

Kidd, D. C. and E. Castano (2013), "Reading literary fiction improves theory of mind," *Science*, 342:377–80.

Kim, J., S. LeBlanc, S. Muralidhar, S. Tonegawa, and X. Zhang (2017), "Basolateral to central amygdala neural circuits for appetitive behaviors," *Neuron*, 93:1464–79.

Kinderman, P., R. P. Bentall, and R. Dunbar (1998), "Theory-of-mind deficits and causal attribution," *British Journal of Psychology*, 89:191–204.

Kitamura, T., M. Morrissey, S. Ogawa, T. Okuyama, R. Redondo, D. Roy, L. M. Smith, and S. Tonegawa (2017), "Engrams and circuits crucial for systems consolidation of a memory," *Science*, 356:73–8.

Knausgaard, K. O. (2015), "The terrible beauty of brain surgery," *NY Times Sunday Magazine*, December 30.

Koch, C. (2012), *Consciousness*, Cambridge MA: MIT Press.

Kojima, S., M. Ceugniet, and A. Izumi (2003), "Identification of vocalizers by pant hoots, pant grunts and screams in a chimpanzee," *Primates*, 44:225–30.

Konkle, T. and A. Caramazza (2013), "Tripartite organization of the ventral stream by animacy and object size," *J. Neurosci.*, 33:10235–42.

Konorski, J. (1967), *Integrative Activity of the Brain*, Chicago IL: University of Chicago Press.

Kovacs, A. M., A. D. Endress, and E. Teglas (2010), "The social sense: Susceptibility to others' beliefs in human infants and adults," *Science*, 330:1830–4.

Kragel, P. A., A. R. Hariri, A. R. Knodt, and K. S. LaBar (2016), "Decoding spontaneous emotional states in the human brain," *PLoS Bio.*, 14(9):e2000106.

Krauss, L. M. (2009), "An update on C. P. Snow's 'Two Cultures'," *Scientific American*, August.

Kreitewolf, J., A. D. Friederici, and K. Kriegstein (2014), "Hemispheric lateralization of linguistic prosody recognition in comparison to speech and speaker recognition," *Neuroimage*, 102(pt 2):332–44.

Kuhl, P. (2010), "Brain mechanisms in early language acquisition," *Neuron*, 67:713–27.

Kuhl, P. (2015), "How babies learn language," *Scientific American*, 313(5):64–9.

Lacey, S., K. Sathian, and R. Stilla (2012), "Metaphorically feeling: Comprehending textural metaphors activates somatosensory cortex," *Brain Lang.*, 120:416–21.

Lai, V. T., P. Hagoort, and R. M. Willems (2015), "Feel between the lines: Implied emotion in sentence comprehension," *J. Cognitive Neurosci.*, 27:1528–41.

Lakoff, G. and M. Johnson (1980), *Metaphors We Live By*, Chicago IL: University of Chicago Press.

Lashley, K. S. (1950), "In search of the engram," *Symposiums of the Society of Experimental Biology*, 4:454–82.

Leavis, F. R. (1962), "Two Cultures? The Significance of C.P. Snow," *The Spectator*, 9 March.

LeDoux, J. (1996), *The Emotional Brain: The Mysterious Underpinnings of Emotional Life*, New York: Simon and Schuster.

LeDoux, J. (2017), "Semantics, surplus meaning, and the science of fear," *Trends Cognitive Sci.*, 21:303–6.

LeDoux, J. and A. Damasio (2013), "Emotions and feelings," in E. Kandel, A. J. Hudspeth, Thomas M. Jessell, James H. Schwartz, and Steven A. Siegelbaum (eds), *Principles of Neural Science*, 5th edn, chapter 48, New York: McGraw-Hill Companies, inc.

Leech, G. N. and M. H. Short ([1981] 2007), *Style in Fiction*, London: Longman.

LePort, A. R., L. Cahill, H. Dickinson-Anson, J. H. Fallon, F. Kruggel, A. T. Mattfield, J. L. McGaugh, and C. Stark (2012), "Behavioral and neuroanatomical investigation of highly superior autobiographical memory (HSAM)," *Neurobiol. Learning & Memory*, 98:78–92.

Lettvin, J. Y., H. R. Maturana, W. S. McCulloch, and W. H. Pitts (1959), "What the frog's eye tells the frog's brain," *Proceedings Inst. Radio Engr.*, 47:1940–51.

Lewin, R. (1982), "Biology is not postage stamp collecting: Interview with Ernst Mayr," *Science*, 216:718–20.

Lewis, P. A., R. Brown, R. I. M. Dunbar, R. Rezaie, and N. Roberts (2011), "Ventromedial prefrontal volume predicts understanding of others and social network size," *Neuroimage*, 57:1624–9.

Lewis, P. A., A. Birch, R. I. M. Dunbar, and A. Hall (2017), "Higher order intentionality tasks are cognitively more demanding," *Social Cognitive Affect. Neurosci.*, 12:1063–71.

Lin, N., H. Hua, J. Li, X. Li, Y. Ma, S. Wang, and X. Wang (2018), "Neural correlates of three cognitive processes involved in theory of mind and discourse comprehension," *Cognitive Affect. & Behav. Neurosci.*, 18:273–83.

Lockwood, P. and M. Wittmann (2018), "Ventral anterior cingulate cortex and social decision-making," *Neurosci. & Biobehavioral Revs.*, 92:187–91.

Lodge, D. (1966), *Language of Fiction*, London: Routledge and Keegan Paul.

Lodge, D. (2002), "Consciousness and the novel," in *Consciousness and the Novel: Connected Essays*, chapter 1, Cambridge MA: Harvard University Press.

Luria, A. (1987), *Mind of a Mnemonist*, Cambridge MA: Harvard University Press.

Magnee, M., B. de Gelder, C. Kemner, and J. Stekelenburg (2007), "Similar facial electromyographic responses to faces, voices, and body expressions," *Neuroreport*, 18:369–72.

Maguire, E. A., H. J. Spiers, C. D. Good, T. Hartley, R. S. J. Frackowiak and N. Burgess (2003), "Navigation expertise and the human hippocampus: A structural brain imaging analysis," *Hippocampus* 13:250–9.

Mantini, D. and W. Vanduffel (2012), "Emerging roles of the brain's default network," *The Neuroscientist*, 19:76–87.

Mar, R. (2004), "The neuropsychology of narrative: Story comprehension, story production and their interrelation," *Neuropsychologia*, 42:1414–34.

Mar, R. (2011), "The neural bases of social cognition and story comprehension," *Annual Rev. Psychology*, 62:103–34.

Mar, R. A., J. de la Paz, J. Hirsh, K. Oatley, and J. B. Peterson (2006), "Bookworms versus nerds: Exposure to fiction versus non-fiction, divergent associations with social ability, and the simulation of fictional social worlds," *J. Res. Personality*, 40:694–712.

Marinkovic, K. (2004), "Spatiotemporal dynamics of word processing in the human cortex," *The Neuroscientist*, 10:142–52.

Marinkovic, K., V. Carr, A.M. Dale, R. P. Dhond, M. Glessner, and E. Halgren (2003), "Spatiotemporal dynamics of modality-specific and supramodal word processing," *Neuron*, 38:487–97.

Marinkovic, K., S. Baldwin, M. G. Courtney, A. M. Dale, E. Halgren, and T. Witzel (2011), "Right hemisphere has the last laugh: Neural dynamics of joke appreciation," *Cognitive Affect. Behav. Neurosci.*, 11(1):113–30.

Marsh, H. (2015), *Do No Harm: Stories of Life, Death, and Brain Surgery*, New York: Thomas Dunne Books (St. Martin's Press).

Marsh, H. (2018), *Admissions: Life as a Brain Surgeon*, New York: Picador.

Martin, A., J. V. Haxby, L. G. Ungerleider, and C. L. Wiggs (1996), "Neural correlates of category-specific knowledge," *Nature*, 379:649–52.

Martin, C. B., J. G. Burneo, B. Hayman-Abello, S. Köhler, S. M. Mirsattari, S. Pietrantonio, and J. C.Pruessner (2012), "Déjà vu in unilateral temporal lobe epilepsy is associated with selective familiarity impairments on experimental tasks of recognition memory," *Neuropsychologia*, 50:2981–91.

Martin, S., P. Brunner, N. E. Crone, H. J. Heinze, C. Holdgraf, R. T. Knight, B. N. Pasley, J. Rieger, and G. Schalk (2014), "Decoding spectrotemporal features of overt and covert speech from the human cortex," *Frontiers Neuroeng.*, 7:14.

Martinez, L., H. Aviezer, V. Falvello, and A. Todorov (2016), "Contributions of facial expressions and body language to the rapid perception of dynamic emotions," *Cognition & Emotion*, 30:939–52.

Mashal, N., M. Faust, T. Hendler, and M. Jung-Beeman (2007), "An fMRI investigation of the neural correlates underlying the processing of novel metaphoric expressions," *Brain Lang.*, 100(2):115–26.

Massaro, D., C. Di Dio, D. Freedberg, V. Gallese, G. Gilli, A. Marchetti and F. Savazzi (2012), "When art moves the eyes: A behavioral and eye-tracking study," *PLoS One*, 7:e37285.

Masuda, K., Ishii, K. Ito, and T. H. Wang (2012), "Do surrounding figures' emotions affect judgement of the target figure's emotion? Comparing the eye-movement patterns of European Canadians, Asian Canadians, Asian international students, and Japanese," *Frontiers Integrative Neurosci.*, 6:article 72.

Masuda, T., P. C. Ellsworth, J. Leu, B. Mesquita, S. Tanida, and E. Van de Veerdonk (2008), "Placing the face in context: Cultural differences in the perception of facial emotion," *J. Personality & Social Psychol.*, 94(3):365–81.

Matsuda, Y.- T., K. Cheng, T. Fujimura, K. Katahira, M. Okada, K. Okanoya, and K. Ueno (2013), "The implicit processing of categorical and dimensional strategies: An fMRI study of facial emotion perception," *Frontiers Human Neurosci.*, 7(551).

McConachie, B. (2007), "Falsifiable theories for theatre and performance studies," *Theatre Journal*, 59(4):December.

McEwan, I. (1987), *The Child in Time*, London: Jonathan Cape.

McEwan, I. (1997), *Enduring Love*, London: Jonathan Cape.

McEwan, I. (2005a), *Saturday: A Novel*, New York: Nan A. Talese (division of Random House, Inc.).

McEwan, I. (2005b), "Literature science and human nature," in J. Gottschall and D.S. Wilson (eds), *The Literary Animal*, 5–19, Evanston IL: Northwestern University Press.

McEwan, I. (2019), *Machines Like Me*, London: Jonathan Cape.

McGilchrist, I. (2009), *The Master and His Emissary: The Divided Brain and the Making of the Western World*. New Haven CT: Yale University Press.

McGrone, J. (1999), *Going Inside*, London: Faber & Faber.

Medaglia, J. D., D. S. Bassett, and M. E. Lynall (2015), "Cognitive network neuroscience," *J. Cognitive Neurosci.*, 27(8):1471–91.

Mendelsund, P. (2014), *What We See When We Read*, New York: Vintage.

Mercier, H. and D. Sperber (2017), *The Enigma of Reason*, Cambridge MA: Harvard University Press.

Mearns, William Hughes (1955), *What Cheer, an Anthology of American and British Humorous and Witty Verse*, New York: Modern Library, 429.

Miall, D. (1989), "Beyond the schema given: Affective comprehension of literary narratives," *Cognition & Emotion*, 3:55–78.

Milton, J. ([1667] 2011), *Paradise Lost*, London: Taylor & Francis.

Mithen, S. (1996), *The Prehistory of the Mind: The Cognitive Origins of Art and Science*, London: Thames and Hudson.

Moerel, M., F. De Martino, and E. Formisano (2012), "Processing of natural sounds in human auditory cortex: Tonotopy, spectral tuning and relation to voice sensitivity," *J. Neurosci.*, 32:14205–14216.

Molenberghs, P., J. D. Henry, H. Johnson, and J. B. Mattingly (2016), "Understanding the minds of others: A neuroimaging meta-analysis," *Neurosci. Biobehavioral Rev.*, 65:276–91.

Morissey, M., N. Insel, and K. Takehara-Nishiuchi (2017), "Generalizable knowledge outweighs incidental details in prefrontal ensemble code over time," *eLIFE*, 6:22177.

Morris, J. S., R. J. Dolan, and A. Ohman (1998), "Conscious and unconscious emotional learning in the human amygdala," *Nature*, 393:467–70.

Moscovitch, M., R. Cabeza, L. Nadel, and G. Winocur (2016), "Episodic memory and beyond: The hippocampus and neocortex in transformation," *Annual Rev. Psychology*, 67:105–34.

Moseley, R. and F. Pulvermüller (2014), "Nouns, verbs, objects, actions, and abstractions: Local fMRI activity indexes semantics, not lexical categories," *Brain & Language*, 132:28–42.

Mullally, S. and E. Maguire (2014), "Memory, imagination and predicting the future: A common brain mechanism?" *The Neuroscientist*, 20:220–34.

Neisser, U. (2008), "Memory with a grain of salt," in H. Harvey Wood and A. S. Byatt (eds), *Memory, An Anthology*, 80–8, London: Chatto & Windus.

Nell, V. (1988), "The psychology of reading for pleasure: Needs and gratifications," *Reading Research Quarterly* 23:6–50.

Noe, A. (2011), "Art and the limits of neuroscience," *New York Times*, December 4.

Norenzayan, A. and S. J. Heine (2005), "Psychological universals: What are they and how can we know?" *Psychol. Bulletin*, 131:763–84.

Norman-Haignere, S., N. G. Kanwisher, and J. H. McDermott (2015), "Distinct cortical pathways for music and speech revealed by hypothesis free voxel decomposition," *Neuron*, 88:1281–96.

Northoff, G. and A. Heinzel (2006), "First-person neuroscience: A new methodological approach for linking mental and neuronal states," *Philosophy, Ethics & Humanities in Medicine*, 1:3.

Nottebohm, F. (1981), "A brain for all seasons: Cyclical anatomical changes in song control nuclei of the canary brain," *Science*, 214:1368–70.

Oates, J. C. ([2016] 2017), *The Man Without a Shadow*, New York: Ecco, HarperCollins.

Oatley, K. (2011), *Such Stuff as Dreams: The Psychology of Fiction*, West Sussex UK: Wiley-Blackwell.

Oatley, K. and P. Johnson-Laird (2014), "Cognitive approaches to emotions," *Trends Cognitive Sci.*, 18:134–40.

Ojemann, G. (1991), "Cortical organization of language," *J. Neurosci.*, 11:2281–7.

Olsson, A. and E. A. Phelps (2007), "Social learning of fear," *Nature Neurosci.*, 10:1095–102.

Ortega, F. and F. Vidal (2013), "Brains in literature/literature in the brain," *Poetics Today*, 34(3):327–60.

Ortony, A. and T. J. Turner (1990), "What's basic about basic emotions?" *Psychol. Rev.*, 97:315–31.

O'Shea, M. (2006), *The Brain: A Very Short Introduction*, Oxford UK: Oxford University Press.

Owen, M. (2017), *Neuroscience, Consciousness and Neurofiction*, Ph.D thesis, University of British Columbia, Vancouver.

Oya, H., R. Adolphs, M. A. Howard, and H. Kawasaki (2002), "Electrophysiological responses in the human amygdala discriminate emotion categories of complex stimuli," *J. Neurosci.*, 22:9502–12.

Pagel, M. (2016), "Anthropology: The long lives of fairy tales," *Current Biol.*, 26:R279–R281.

Palmer, A. (2005), "Thought in the novel: The middlemarch mind," *Style*, 34(4):427–39.

Palmer, A. (2010), "Storyworlds and Groups," in L. Zunshine (ed), *Introduction to Cognitive Cultural Studies*, 176–92, Baltimore MD: Johns Hopkins University Press.

Palombo, D. J., C. Alain, W. Khuu, B. Levine, and H. Söderlund (2015), "Severely deficient autobiographical memory (SDAM) in healthy adults: A new mnemonic syndrome," *Neuropsychologia*, 72:105–18.

Panero, M. E., D. S. Weisberg, J. Black, T. R. Goldstein, J. L. Barnes, H. Brownell, and E. Winner (2016), "Does reading a single passage of literary fiction really improve theory of mind? An attempt at replication," *J. Personality Soc. Psychology* 111(5):e46–e54.

Panksepp, J. (1998), *Affective Neuroscience*, Oxford UK: Oxford University Press.

Panksepp, J. (2011), "The basic emotional circuits of mammalian brains: Do animals have affective lives?" *Neurosci. & Biobehavioral Revs.*, 35:1791–804.

Pasley, B. N., E. F. Chang, N. E. Crone, S. V. David, A. Finker, R. T. Knight, N. Mesgarani, and S. A. Shamma (2012), "Reconstructing speech from human auditory cortex, *PLoS Biol.*, 10:e1001251.

Pearson, J. (2019), "The human imagination: The cognitive neuroscience of visual mental imagery," *Nature Revs. Neuroscience* 20(10):624–34.

Pearson, J. and S. M. Kosslyn (2015), "The heterogeneity of mental representation: Ending the imagery debate," *Proceedings Nat. Acad. Sci.*, 112:10089–9992.

Pegado, F., L. Cohen, S. Dehaene, and K. Nakamura (2011), "Breaking the symmetry: Mirror discrimination for single letters but not for pictures in the visual word form area," *Neuroimage*, 55(2):742–9.

Pegado, F., L. W. Braga, M. Buiatti, L. Cohen, E. Comerlato, S. Dehaene, G. Dehaene-Lambertz, A. Jobert, R. Kolinsky, J. Morais, K. Nakamura, F. Ventura, and P. Ventura (2014), "Timing the impact of literacy on visual processing," *Proceedings Nat. Acad. Sci.*, 111(49):E5233–E5242.

Penfield, W. and T. Rasmussen (1950), *The Cerebral Cortex of Man: A Clinical Study of Localization of Function*, New York: Macmillan.

Perrachione, T. K., S. N. Del Tufo and J. D. E. Gabrieli (2011), "Human Voice Recognition Depends on Language Ability," *Science* 333:595.

Petersen, S. E. and O. Sporns (2015), "Brain networks and cognitive architectures," *Neuron*, 88:207–19.

Petersen, S. E., J. A. Fiez, M. E. Raichle, and H. van Mier (1998), "The effects of practice on the functional anatomy of task performance," *Proceedings Nat. Acad. Sci.*, 95:853–60.

Peterson, D. (2004), "Science, scientism, and professional responsibility," *Clinical Psychol.: Sci. Practice.*, 11:196–210.

Phelps, E. A. (2006), "Emotion and cognition: Insights from studies of the human amygdala," *Annual Rev. Psychol.*, 57:27–53.

Pinker, S. (2010), "The Cognitive Niche: Coevolution of intelligence, sociality, and language," *Proceedings Nat. Acad. Sci.*, 107:8993–9.

Pisanski, K., V. Cartei, C. McGettigan, J. Raine, and D. Reby (2016), "Voice modulation: A window into the origins of human vocal control?" *Trends Cognitive Sci.*, 20:304–18.

Pobric, G., M. Faust, M. Lavidor, and N. Mashal (2008), "The role of the right cerebral hemisphere in processing novel metaphoric expressions: A transcranial magnetic stimulation study," *J. Cognitive Neurosci.*, 20(1):170–81.

Pobric, G., E. Jeffries, M. A. Lambon Ralph (2010), "Amodal semantic representations depend on both anterior temporal lobes: Evidence from repetitive transcranial magnetic stimulation," *Neuropsychologia*, 48:1336–42.

Poeppel, D. (2014), "The neuroanatomic and neurophysiological infrastructure for speech and language," *Current Opinions Neurobiol.*, 28:142–9.

Poole, S. (2012), "Your brain on pseudoscience: The rise of popular neurobollocks," *New Statesman*, September 6.

Powell, J. L., R. I. M. Dunbar, M. García-Fiñana, P. A. Lewis, and N. Roberts (2010), "Orbital prefrontal cortex volume correlates with social cognitive competence," *Neuropsychologia* 48:3554–62.

Powell, J. L., R. I. M. Dunbar, M. García-Fiñana, P. A. Lewis, and N. Roberts (2014), "Orbital prefrontal cortex volume predicts social network size: An imaging study of individual differences in humans," *Proceedings Royal Society Lond.*, B279:2157–62.

Premack, D. and G. Woodruff (1978), "Does the chimpanzee have a theory of mind?" *Behavioral Brain Sci.*, 1:515–26.

Pulvermüller, F. (2013), "How neurons make meaning: Brain mechanisms for embodied and abstract-symbolic semantics," *Trends Cognitive Sci.*, 17:458–70.

Pulvermüller, F. and L. Fadiga (2010), "Active perception: Sensorimotor circuits as a cortical basis for language," *Nature Revs. Neurosci.*, 11:351–60.

Pulvermüller, F., O. Hauk, F. Kherif, B. Mohr, and I. Nimmo-Smith (2009), "Distributed cell assemblies for general lexical and category specific semantic processing as revealed by fMRI cluster analysis," *Human Brain Mapping*, 30(12):3837–50.

Quian Quiroga, R. (2012a), *Borges and Memory: Encounters with the Human Brain*, Cambridge MA: MIT Press.

Quian Quiroga, R. (2012b), "Concept cells: The building blocks of declarative memory functions," *Nature Revs. Neurosci.*, 13:587–97.

Quian Quiroga, R., I. Fried, C. Koch, and A. Kraskov (2009), "Explicit encoding of multimodal percepts by single neurons in the human brain," *Current Biol.*, 19:130813.

Rachul, C. and A. Zarzeczny (2012), "The rise of Neuroskepticism," *International Journal of Law and Psychiatry*, 35(2):77–81.

Raichle, M. (1994), "Visualizing the mind," *Scientific American*, 270(4):58–64.

Raichle, M. (2006), "The brain's dark energy," *Science*, 314:1249–50.

Raichle, M. (2008), "A brief history of human brain mapping," *Trends Neurosci.*, 32:118–26.

Raichle, M. (2011), "The restless brain," *Brain Connect.*, 1(1):3–12.

Raichle, M. (2015), "The brain's default mode network," *Annual Rev. Neurosci.*, 38:433–47.

Raine, J., A. Oleszkiewicz, K. Pisanski, D. Reby, and A. Simner (2018), "Human listeners can accurately judge strength and height relative to self from aggressive roars and speech," *iScience*, 4:273–80.

Ramus, F. (2004), "Neurobiology of dyslexia: A reinterpretation of the data," *Trends Neurosci.*, 27:720–6.

Rayner, K. (1998), "Eye movements in reading and information processing: 20 years of research," *Psychol. Bulletin*, 124:372–422.

Rayner, K. and E. D. Reichle (2010), "Models of the reading process," *WIREs Cognitive Science*, 1:787–99.

Reddy, L. and N. Kanwisher (2006), "Coding of visual objects in the ventral stream," *Current Opinion Neurobiol.*, 16:408–14.

Regev, M., U. Hasson, C. J. Honey, and E. Simony (2013), "Selective and invariant neural responses to spoken and written narratives," *J. Neurosci.*, 33:15978–88.

Rendu, W., A. Balzeau, P. Bayle, C. Beauval, T. Bismuth, L. Bourguignon, I. Crevecoeur, G. Delfour, J. P. Faivre, F. Lacrampe-Cuyaubère, B. Maureille, C. Tavormina, D. Todisco, and A. Turq (2014), "Evidence supporting an intentional Neanderthal burial at La Chapelle-aux-Saints," *Proceedings Nat. Acad. Sci.*, 111:81–6.

Rice, H. J. and D. C. Rubin (2009), "I can see it both ways: First- and third-person visual perspectives at retrieval," *Consciousness & Cognition*, 18:877–90.

Rice, H. J. and D. C. Rubin (2011), "Remembering from any angle: The flexibility of visual perspective during retrieval," *Consciousness & Cognition*, 20:568–77.

Richards, I. A. ([1924] 2001), *Principles of Literary Criticism*, London: Routledge.

Richards, I. A. ([1934] 1966), *Coleridge on the Imagination*, Bloomington IN: Indiana University Press.

Richardson, A. (2001), *British Romanticism and the Science of the Mind*, Cambridge UK: Cambridge University Press.

Richardson, A. (2010a), "The romantic image, the mind's eye, and the history of the senses," in *The Neural Sublime: Cognitive Theories and Romantic Texts*, Baltimore MD: Johns Hopkins University Press.

Richardson, A. (2010b), "Facial expression theory from romanticism to the present," in L. Zunshine (ed), *Introduction to Cognitive Cultural Studies*, 65–83, Baltimore MD: Johns Hopkins University Press.

Richardson, A. (2011), "Memory and imagination in romantic fiction," in S. Nalbantian, P. Matthews, and J. McClelland (eds), *The Memory Process: Neuroscientific and Humanistic Perspectives*, chapter 13, Cambridge MA: MIT Press.

Rimmon-Kenan, S. (2002), *Narrative Fiction: Contemporary Poetics*, London: Routledge.

Rimmon-Kenan, S. (2015), *A Glance beyond Doubt: Narration, Representation, Subjectivity*, New York: Ohio University Press.

Rizzolatti, G., L. Fadiga, L. Fogassi, and V. Gallese (1996), "Premotor cortex and the recognition of motor actions," *Cognitive Brain Res.*, 3:131–41.

Roche King, J. and C. Comer (1996), "Visually elicited turning behavior in Rana pipiens: comparative organization and neural control of escape and prey capture," *J. Comp. Physiol A*, 178:293–305.

Roeder, K. D. (1959), "A physiological approach to the relation between prey and predator," *Smithson. Misc. Coll.*, 137:287–306.

Rogers, J. (2014), *Unified Fields*, Montreal: McGill University Press.

Rolls, E. T. and S. Wirth (2018), "Spatial representations in the primate hippocampus, and their functions in memory and navigation," *Progress in Neurobiol.*, 171:90–113.

Root-Bernstein, R. (1997), "Macroscope: Art, Imagination and the Scientist," *American Scientist*, 85(1):6–9.

Rorden, C. and H. O. Karnath (2004), "Using human brain lesions to infer function: A relic from a past era in the fMRI age?" *Nature Revs. Neurosci.*, 5:813–19.

Roy, C. S. and C. S. Sherrington (1890), "On the regulation of the blood supply of the brain," *J. Physiology* 11:85–108.

Russell, B. ([1927] 1960), *An Outline of Philosophy*, New York: Meridien – World Publishing.

Russell, J. (2003), "Core affect and the psychological construction of emotion," *Psychol. Rev.*, 110:145–72.

Russo, J. P. (1989), *I. A. Richards: His Life and Work*, New York: Routledge.

Ryan, M.- L. (1980), "Fiction, non-factuals, and the principle of minimal departure," *Poetics*, 9:403–22.

Ryan, M.- L. (2010), "Narratology and cognitive science: A problematic relation," *Style*, 44:469–95.

Ryan, M. L., K. Foote, and M. Azaryahu (2016), *Narratizing Space/Spatializing Narrative*, Columbia: Ohio State University Press.

Sagan, C. ([1977] 1978), *The Dragons of Eden: Speculations on the Evolution of Human Intelligence*, New York: Ballantine Books.

Satel, S. and S. Lillienfeld (2013), *Brainwashed: The Seductive Appeal of Mindless Neuroscience*, New York: Basic Books.

Sato, T., J. Kitazono, M. D. Lescroart, M. Okada, M. Tanifuji, and G. Uchida (2013), "Object representation in inferior temporal cortex is organized in a mosaic-like structure," *J. Neurosci.*, 33(42):16642–56.

Sauter, D. and F. Eisner (2013), "Commonalities outweigh differences in the communication of emotions across human cultures," *Proceedings Nat. Acad. Sciences*, 110(3):E180.

Scarry, E. ([1999] 2001), *Dreaming by the Book*, Princeton NJ: Princeton University Press.

Schacter, D. L. and D. R. Addis (2007), "The ghosts of past and future," *Nature*, 445:27.

Schaefer, M., H. J. Heinze, Y. Rothemund, and M. Rotte (2004), "Short-term plasticity of the primary somatosensory cortex during tool use," *Neuroreport*, 15:1293–7.

Schank, R. C. (1995), *Tell Me a Story: Narrative and Intelligence*, Evanston IL: Northwestern University Press.

Schechtman, M. ([1996] 2007), *The Constitution of Selves*, Ithaca NY: Cornell University Press.

Schmandt-Besserat, D. (1981), "Decipherment of the earliest tablets," *Science*, 211:283–5.

Schmandt-Besserat, D. (2007), *When Writing Met Art: From Symbol to Story*, Austin TX: University of Texas Press.

Schneider, G. E. (1969), "Two visual systems," *Science*, 163:895–902.

Schweinkreis, P., H. R. Dinse, S. El Tom, B. Pleger, P. Ragert, and M. Tegenthoff (2007), "Assessment of sensorimotor cortical representation asymmetries and motor skills in violin players," *Eur. J. Neurosci.*, 26:3291–302.

Searle, J. (1969), *Speech Acts*, Cambridge UK: Cambridge University Press.

Sereno, M. I. and R. S. Huang (2014), "Multisensory maps in parietal cortex," *Current Opinions Neurobiol.*, 24:39–46.

Seyfarth, R. M. and D. L. Cheney (1992), "Meaning and mind in monkeys," *Scientific American*, 267(6):122–8.

Shainberg, L. (1979), *Brain Surgeon: An Intimate View of His World*, New York: J.B. Lippincott Company.

Shainberg, L. (1987), "Exorcising Beckett," *Paris Review*, 104.

Shainberg, L. (1988), *Memories of Amnesia: A Novel*, Latham NY: Paris Review Editions, British American Publishing Ltd.

Shears, C., A. Hawkins, A. Varner, L. Lewis, J. Heatley, and L. Twachtmann (2008), "Knowledge-based inferences across the hemispheres: Domain makes a difference," *Neuropsychologia* 46:2563–8.

Sherratt, S. and K. Bryan (2012), "Discourse production after right brain damage: Gaining a comprehensive picture using a multi-level processing model," *J. Neurolinguistics*, 25:213–39.

Sherrington, C. S. (1906), *The Integrative Action of the Nervous System*, New Haven CT: Yale University Press.

Sherrington, C. S. ([1940] 2009), *Man on His Nature*, Cambridge UK: Cambridge University Press.

Signoret, J. L., R. Abelanet, P. Castaigne, P. Lavorel, and F. Lhermitte (1984), "Rediscovery of Leborgne's brain: Anatomical description with CT scan," *Brain and Language*, 22:303–19.

Silbert, L. J., U. Hasson, C. J. Honey, D. Poeppel, and E. Simony (2014), "Coupled neural systems underlie the production and comprehension of naturalistic narrative speech," *Proceedings Nat. Acad. Sci.*, 111:E4687–E4696.

Silk, J. B. (2007), "Social components of fitness in primate groups," *Science*, 317:1347–51.

Silk, J. B. and B. R. House (2011), "Evolutionary foundations of human prosocial sentiments," *Proceedings Nat. Acad. Sci.*, 108:10910–17.

Silk, J. B., S. C. Alberts, and J. Altmann (2003), "Social bonds of female baboons enhance infant survival," *Science*, 302:1231–34.

Silk, J. B., J. C.Beehner, T. J. Bergman, D. L. Cheney, C. Crockford, A. L. Engh, L. R. Moscovice, R. M. Seyfarth, and R. M. Wittig (2009), "The benefits of social capital: Close social bonds among female baboons enhance survival of offspring," *Proceedings Royal Society Lond.*, B276:3099–104.

Simon, H. and C. Kaplan ([1989] 1993), "Foundations of cognitive science," in M. I. Posner (ed), *Foundations of Cognitive Science*, Cambridge MA: MIT Press.

Simoniti, C. N., J. M. Akey, L. Bastarache, E. Bottinger, W. S. Bush, J. A. Capra, D. S. Carrell, R. L. Chisholm, D. R. Crosslin, J. C. Denny, S. J. Hebbring, G. P. Jarvik, I. J. Kullo, R. Li, J. Pathak, J. D. Prato, M. D. Ritchie, D. M. Roden, G. Tromp, S. S. Verma, and B. Vernot (2016), "The phenotypic legacy of admixture between modern humans and Neanderthals," *Science*, 351:737–41.

Simonyan, K., H. Ackermann, E. F. Chang, and J. D. Greenlee (2016), "New developments in understanding the complexity of human speech production," *J. Neurosci.* 36:11440–8.

Siri, F., M. Ardizzi, M. Becaria, C. Christov-Bakargiev, F. Feroni, V. Gallese, A. Kolesnikova, and B. Rocci (2018), "Behavioral and autonomic responses to real and digital reproductions of art," *Progress in Brain Research*, 237:201–21.

Slade, E. S. (1997), *Analysing Casual Conversation*, London: Equinox.

Slotnick, S. D., S. M. Kosslyn, and W. L. Thompson (2005), "Visual mental imagery induces retinotopically organized activation of early visual areas," *Cereb. Cortex*, 15:1570–83.

Smallwood, J., B. Baird, H. Engen, J. Golchert, K. J. Gorgolewski, D. S. Margulies, F. J. Ruby, J. W. Schooler, and M. T. Vinski (2013), "The default modes of reading: Modulation of posterior cingulate and medial prefrontal cortex

connectivity associated with comprehension and task focus while reading,"
Frontiers Human Neurosci., 7:734.

Smith, D., L. Astete, N. Chaudhary, M. Dyble, R. Mace, K. Major, M. Ngales,
A. E. Page, G. D. Salali, P. Schlaepfer, J. Thompson, and L. Vinicius (2017),
"Cooperation and the evolution of hunter-gatherer storytelling," *Nature
Commun.*, 8:1853–61.

Snow, C. P. ([1959] 1998), *The Two Cultures*, Canto edn, Cambridge UK:
Cambridge University Press.

Snyder, A. Z. and M. Raichle (2012), "A brief history of the resting state: The
Washington University perspective," *Neuroimage*, 62:902–10.

Speer, N., J. Reynolds, and J. Zacks (2007), "Human brain activity time-locked to
narrative event boundaries," *Psychological Sciences*, 18:449–55.

Spolsky, E. (1993), *Gaps in Nature: Literary Interpretation and the Modular Mind*,
Albany NY: State University of New York Press.

Srihasam, K., M. S. Livingstone, J. B. Mandeville, I. A. Morocz, and K. J. Sullivan
(2012), "Behavioral and anatomical consequences of early versus late symbol
training in macaques," *Neuron*, 73(3):608–19.

Srihasam, K., M. S. Livingstone, and J. L. Vincent (2014), "Novel domain
formation reveals proto-architecture in inferotemporal cortex," *Nature
Neurosci.*, 17:1776–85.

Stanley, D. A. and R. Adolphs (2013), "Toward a neural basis for social behavior,"
Neuron, 80:816–26.

Starr, G. G. (2013), *Feeling Beauty: The Neuroscience of Aesthetic Experience*,
Cambridge MA: MIT Press.

Steadman, G. (2008), "Brain plots: Neuroscience and the contemporary novel,"
Yearbook of Research in English and American Literature, 24:113–24.

Stein, B. E. and M. A. Meredith (1993), *The Merging of the Senses*, Cambridge
MA: MIT Press.

Stephens, G. J., U. Hasson, and L. J. Silbert (2010), "Speaker-listener neural
coupling underlies successful communication," *Proceedings Nat. Acad. Sci.*,
107:14425–30.

Stiles, A. (2007), *Neurology and Literature, 1860–1920*, New York: Palgrave
Macmillan.

Stockwell, P. (2009), *Texture: A Cognitive Aesthetics of Reading*, Edinburgh:
Edinburgh University Press.

Stokes, M., R. Cusack, J. Duncan, and R. Thompson (2009), "Top-down activation
of shape-specific population codes in visual cortex during mental imagery,"
J. Neurosci., 29(5):1565–72.

Stokes, R., G. Hickok, and J. Venezia (2019), "The motor system's [modest]
contribution to speech perception," *Psychonomic Bulletin & Rev.*, 26:1354–66.

Strawson, G. (2015), "I am not a story," *Aeon*. September 3. Available online:
https://aeon.co/essays/let-s-ditch-the-dangerous-idea-that-life-is-a-story

Striedter, G. F. (2005), *Principles of Brain Evolution*, Sunderland MA: Sinauer
Associates.

Suga, N. (1989), "Principles of auditory information processing derived from
neuroethology," *J. Exp. Biol.*, 146:277–86.

Suthana, N. and I. Fried (2012), "Percepts to recollections: Insights from single
neuron recordings in the human brain," *Trends Cognitive Sci.*, 16:427–36.

Swanson, F. J., C. Goodrich, and K. D. Moore (2008), "Bridging boundaries: Scientists, creative writers, and the long view of the forest," *Frontiers Ecol. Environ.*, 6:499–504.

Szpunar, K. K., K. B. McDermott, and J. M. Watson (2007), "Neural substrates of envisioning the future," *Proceedings Nat. Acad. Sci.*, 104:642–7.

Taleb, N. N. (2008), *The Black Swan*, London: Penguin.

Tallis, R. (2008), "The neuroscience delusion: Neuroaesthetics is wrong about our experience of literature—and it is wrong about humanity," *The Times Literary Supplement*, April 9.

Tallis, R. (2011), *Aping Mankind: Neuromania, Darwinitis and the Misrepresentation of Humanity*, Durham UK: Acumen Publishing.

Tamir, D. I., A. Bricker, D. Dodell-Feder, and J. Mitchell (2016), "Reading fiction and reading minds: The role of simulation in the default mode network," *Social Cognitive Affect. Neurosci.*, 11:215–24.

Tanaka, K. (2003), "Columns for complex visual object features in the inferotemporal cortex: Clustering of cells with similar but slightly different stimulus selectivities," *Cerebral Cortex*, 13:90–9.

Taylor, G. (2006), "Ricoeur's philosophy of imagination," *J. French Philosophy*, 16:93–104.

Tettamanti, M., G. Buccino, S. F. Cappa, M. Danna, F. Fazio, V. Gallese, D. Perani, G. Rizzolatti, M. C. Saccuman, and P. Scifo (2005), "Listening to action-related sentences activates fronto-parietal motor circuits," *J. Cognitive Neurosci.*, 17(2):273–81.

Thierry, G., P. M. Davis, V. Gonzalez-Diaz, C. D. Martin, R. Rezaie, and N. Roberts (2008), "Event-related potential characterization of the Shakespearean functional shift in narrative sentence structure," *Neuroimage*, 40:923–31.

Tompkins, J. P. (1977), "Criticism and feeling," *College English*, 39:169–78.

Tompkins, J. P. (1980), "The reader in history: the changing shape of literary response," in J. P. Tompkins (ed), *Reader Response Criticism: From Formalism to Post-structuralism*, 201–32, Baltimore: Johns Hopkins University Press.

Tonegawa, S., M. Pignatelli, D. S. Roy, and T. J. Ryan (2015), "Memory engram storage and retrieval," *Current Opinion Neurobiol.*, 35:101–9.

Tooby, J. and I. DeVore (1987), "The reconstruction of hominid behavioral evolution through strategic modeling," in W. G. Kinzey (ed), *The Evolution of Human Behavior: Primate Models*, Albany NY: State University of New York Press.

Tranel, D., R. Adolphs, A. R. Damasio, and H. Damasio (2001), "A neural basis for the retrieval of words for actions," *Cognitive Neuropsychol.*, 18:655–70.

Tranel, D., R. Adolphs, A. R. Damasio, H. Damasio, and D. Kemmerer (2003), "Neural correlates of conceptual knowledge for actions," *Cognitive Neuropsychol.*, 20(3):409–32.

Treffert, D. A. (2009), "The savant syndrome: An extraordinary condition. A synopsis: Past, present, future," *Philosophical Trans. Royal Society Lond.*, B364:1351–7.

Tsao, D. Y., W. A. Freiwald, R. B. Tootell, M. S. Livingstone (2006), "A cortical region consisting entirely of face-selective cells," *Science*, 311:670–4.

Turchi, P. (2004), *Maps of the Imagination: The Writer as Cartographer*, San Antonio TX: Trinity University Press.

Turner, M. (1991), *Reading Minds: The Study of English in the Age of Cognitive Science*, Princeton: Princeton University Press.

Turner, M. (1996), *The Literary Mind: The Origins of Thought and Language*, Oxford UK: Oxford University Press.

Ungerleider, L. and A. Bell (2011), "Uncovering the visual "alphabet": Advances in our understanding of object perception," *Vision Res.*, 51:782–99.

Van Dijk, T. A. and W. Kintsch (1983), *Strategies in Discourse Comprehension*, New York: Academic Press.

Vannuscorps, G. and A. Caramazza (2016), "Typical action perception and interpretation without motor simulation," *Proceedings Nat. Acad. Sci.*, 113:86–91.

Van Peer, W. and S. Chatman, eds (2001), *New Perspectives on Narrative Perspective*, Albany NY: State University of New York Press.

Vermeule, B. (2010), *Why Do We Care about Literary Characters?*, Baltimore MD: Johns Hopkins University Press.

Vezzali, L., S. Stathi, D. Giovannini, D. Capozza, and E. Trifiletti (2015), "The greatest magic of Harry Potter: Reducing prejudice," J. Applied Social Psychology 45:105–21.

Vico, G. ([1709] 1990), *On the Study Methods of Our Time*, trans. E. Gianturco, Ithaca NY: Cornell University Press.

Vogel, A. C., F. M. Miezen, S. E. Petersen, and B. L. Schlaggar (2012), "The putative visual word form area is functionally connected to the dorsal attention network," *Cerebral Cortex*, 22:537–49.

Walczyk, J. J., T. Tcholakian, F. Igou, and A. P. Dixon (2014), "One hundred years of reading research: Successes and missteps of Edmund Burke Huey and other pioneers," *Reading Psychology* 35:601–21.

Walden, G. (2012), "Beware the Fausts of neuroscience," *Standpoint Magazine*, April.

Wang, X., W. M. Jenkins, M. M. Merzenich, and K. Sameshima (1995), "Remodeling of hand representation in adult cortex determined by timing of tactile stimulation," *Nature*, 378:71–5.

West, D. (2013), *I. A. Richards and the Rise of Cognitive Stylistics*, London: Bloomsbury Academic Press.

Whalen, P. J., N. L. Etcoff, M. A. Jenike, M. B. Lee, S. C. McInerney, and S. L. Rauch (1998), "Masked presentations of emotional facial expressions modulate amygdala activity without explicit knowledge," *J. Neurosci.*, 18(1):411–18.

White, H. (1987), *The Content of the Form*, Baltimore MD: Johns Hopkins University Press.

Whiten, A. and D. Erdal (2012), "The human socio-cognitive niche and its evolutionary origins," *Philosophical Trans. Royal Society Lond.*, B367:2119–29.

Wiessner, P. W. (2014), "Embers of society: Firelight talk among the Ju/hoansi bushmen," *Proceedings Nat. Acad. Sci.*, 111:14027–35.

Wilensky, A. E., M. P. Kristensen, J. E. LeDoux, and G. E. Schafe (2006), "Rethinking the fear circuit: The central nucleus of the amygdala is required for the acquisition, consolidation, and expression of Pavlovian fear conditioning," *J. Neurosci.*, 26:12387–96.

Wilson, E. O. (1998), *Consilience: The Unity of Knowledge*, New York: Vintage Press.

Wilson, M. (2002), "Six views of embodied cognition," *Psychonomic Bulletin & Rev.*, 9(4):625–36.

Wimsatt, W. K. and M. C. Beardsley (1954), "The affective fallacy," in *The Verbal Icon: Studies in the Meaning of Poetry*, 21–39, Lexington KY: University of Kentucky Press.

Winner, E. (2019), *How art works: A psychological explanation*, New York: Oxford University Press.

Wise, R. J. S. and R. M. Braga (2014), "Default mode network: The seat of literary creativity?" *Trends Cognitive Sci.*, 18:116–17.

Witteman, J., N. Schiller, and V. Van Heuven (2012), "Hearing feelings: A quantitative meta-analysis on the neuroimaging literature of emotional prosody perception," *Neuropsychologia*, 50:2752–63.

Wolf, M. ([2007] 2008), *Proust and the Squid: The Story and Science of the Reading Brain*, New York: Harper Perennial Edition.

Wood, James (2008), *How Fiction Works*, New York: Picador, Farrar, Straus, Giroux.

Wood, J. (2015), *The Nearest Thing to Life*, Waltham MA: Brandeis University Press.

Woolf, Virginia ([1919] 2002), "Modern fiction," in *The Common Reader*, Harvest edn, New York: Harcourt.

Yang, X. F., J. Bossmann, M. H. Immordino-Yang, M. Jordan, and B. Schiffhauer (2013), "Intrinsic default mode network connectivity predicts spontaneous verbal descriptions of autobiographical memories during social processing," *Frontiers Psychol.*, 3:592.

Young, K. and J. L. Saver (2001), "The neurology of narrative," *Substance*, 30:72–84.

Zacks, J. M., T. Braver, R. L. Buckner, D. Donaldson, J. Ollinger, M. E. Raichle, M. Sheridan, and A. Snyder (2001), "Human brain activity time-locked to perceptual event boundaries," *Nature Neuroscience*, 4:651–5.

Zacks, J. M., N. Speer, and J. R. Reynolds (2009), "Segmentation in Reading and Film Comprehension," *J. Exp. Psychol: General*, 138:307–27.

Zeidman, P. and E. Maguire (2016), "Anterior hippocampus: The anatomy of perception, imagination and episodic memory," *Nature Revs. Neurosci.*, 17:173–82.

Zhao, Y., J. Ding, X. Du, Z. Han, N. Lin, L. Song, R. Sun, and Q. Wang (2017), "Left anterior temporal lobe and bilateral anterior cingulate cortex are semantic hub regions: Evidence from behavior-nodal degree mapping in brain-damaged patients," *J. Neurosci.*, 37:141–51.

Zimmer, C. and D. J. Emlen (2013), *Evolution: Making Sense of Life*, Greenwood Village CO: Roberts and Company Publishers.

Zunshine, L. (2006), *Why We Read Fiction: Theory of Mind and the Novel*, Columbus OH: Ohio University Press.

Zwaan, R. and G. Radvansky (1998), "Situation models in language comprehension and memory," *Psychol. Bulletin*, 123:162–85.

INDEX